MATHEMATICAL

FORMULAS

AND

SCIENTIFIC DATA

MATHEMATICAL
FORMULAS
AND
SCIENTIFIC DATA

A Quick Reference Guide

C. P. KOTHANDARAMAN, PhD

MERCURY LEARNING AND INFORMATION
Dulles, Virginia
Boston, Massachusetts
New Delhi

Publisher: David Pallai
MERCURY LEARNING AND INFORMATION
22841 Quicksilver Drive
Dulles, VA 20166
info@merclearning.com
www.merclearning.com
(800) 232-0223

C. P. Kothandaraman. *Mathematical Formulas and Scientific Data.*
ISBN: 978-1-68392-865-2

Library of Congress Control Number: 2022952304

232425321   Printed on acid-free paper in the United States of America.

CONTENTS

Preface   *vii*

PART I: MATHEMATICAL FORMULAS   **1**

Chapter 1	Algebra	3
Chapter 2	Elementary Geometry	19
Chapter 3	Trigonometry	27
Chapter 4	Analytic Geometry	43
Chapter 5	Differential Calculus	61
Chapter 6	Integral Calculus—I	75
Chapter 7	Integral Calculus—II	87
Chapter 8	Vectors	133

PART II: SCIENTIFIC DATA FOR ENGINEERS   **137**

Chapter 1	SI Units	139
Chapter 2	The Fundamental Mechanical Units	143
Chapter 3	The Fundamental Electrical and Magnetic Units	145
Chapter 4	Relations Between the Systems of Electrical Units	147
Chapter 5	Rationalization of MKS Units	149
Chapter 6	Definitions of Electric and Magnetic Quantities in SI	151
Chapter 7	Definitions in the CGS Electromagnetic System of Units	155
Chapter 8	Heat Units and Definitions	159
Chapter 9	Photometric and Optical Units and Definitions	163
Chapter 10	Acoustical Units and Definitions	165
Chapter 11	Properties of the Elements	167
Chapter 12	Properties of Metallic Solids (at 293 K)	173
Chapter 13	Properties of Non-Metallic Solids (at 293 K)	179

Chapter 14 Properties of Liquids (at 293 K)   185
Chapter 15 Properties of Gases at STP   189
Chapter 16 Mechanical Data   193
Chapter 17 Relative Humidities From Wet-and Dry-Bulb
                Thermometers (Exposed in Standard Screen)   199
Chapter 18 Acoustic Data   203
Chapter 19 Astronomical Data   207
Chapter 20 Terrestrial and Geodetic Data   211
Chapter 21 Radioactivity   217
Chapter 22 Properties of Inorganic Compounds   221
Chapter 23 Properties of Organic Compounds (at 293 K)   229
Chapter 24 The Fundamental Constants   233

PREFACE

This will serve as an excellent reference for in-service professionals in the areas of mathematics, science, and engineering. It will also help students as a quick review for examinations.

Formulas from algebra to vectors are discussed in Part I of the book. Technical terms, theorems, and applicable laws are defined, and their uses are explained. Answers to standard equations in these areas are provided. Equations from straight lines to conics are derived and applications are provided. In the sections on differential and integral calculus, solutions are included for a large number of situations.

Part II is unique in providing numerical data in astronomical, acoustical, optical, and geophysical areas. Properties such as density, specific heat, thermal conductivity, and expansion coefficients are tabulated for a large number of metallic, nonmetallic, organic, and inorganic materials. This information and data will be particularly useful for practicing engineers.

C.P. Kothandaraman

PART I

MATHEMATICAL FORMULAS

ALGEBRA

1   FUNDAMENTAL LAWS

(a) Commutative law: $a + b = b + a$, $ab = ba$.

(b) Associative law: $a + (b + c) = (a + b) + c$, $a(bc) = (ab)c$.

(c) Distributive law: $a(b + c) = ab + ac$.

2   LAWS OF EXPONENTS

(a) $a^x \cdot a^y = a^{x+y}$, $(ab)^x = a^x \cdot b^x$, $(a^x)^y = a^{xy}$.

(b) $a^0 = 1$ if $a \neq 0$, $a^{-x} = \dfrac{1}{a^x}$, $\dfrac{a^x}{a^y} = a^{x-y}$.

(c) $a^{\frac{x}{y}} = \sqrt[y]{a^x}$, $a^{\frac{1}{y}} = \sqrt[y]{a}$.

3   OPERATIONS WITH ZERO

$a - a = 0$, $a \cdot 0 = 0 \cdot a = 0$.

If $a \neq 0$, $\dfrac{0}{a} = 0$, $0^a = 0$, $a^0 = 1$ (division by zero undefined).

4   COMPLEX NUMBERS

A number of the form $a + ib$, where a and b are real.

$$i = \sqrt{-1} = i^2 = -1, \; i^3 = -i, \; i^4 = 1, \; i^5 = i, \text{ etc.}$$

$$a + ib = c + id, \text{ if and only if } a = c, b = d.$$

$$(a + ib) + (c + id) = (a + c) + id\,(b + d).$$

$$(a + ib)\,(c + id) = (ac - bd) + i\,(ad + bc).$$

$$\frac{a + ib}{c + id} = \frac{(a + ib)(c - id)}{(c + id)(c - id)} = \frac{ac + bd}{c^2 + d^2} + \frac{bc - ad}{c^2 + d^2}\,i.$$

5   LAWS OF LOGARITHMS

If M, N, and b are positive numbers and $b \neq 1$ then

(a) $\log_b M\,N = \log_b M + \log_b N,$ **(b)** $\log_b \dfrac{M}{N} = \log_b M - \log_b N,$

(c) $\log_b M^p = p \cdot \log_b M,$ **(d)** $\log_b \sqrt[q]{M} = \dfrac{1}{q} \cdot \log_b M$

(e) $\log_b \left(\dfrac{1}{M}\right) = -\log_b M,$ **(f)** $\log_b b = 1, \quad \log_b 1 = 0$

Change of base of Logarithms $(c \neq 1)$:

$$\log_b M = \log_c M \cdot \log_b c = \frac{\log_c M}{\log_c b}.$$

6   BINOMIAL THEOREM

(n a positive integer).

$$(a + b)^n = a^n + na^{n-1}b + \frac{n(n-1)}{2!}a^{n-2}b^2$$

$$+ \frac{n(n-1)(n-2)}{3!}a^{n-3}b^3 + \ldots + nab^{n-1} + b^n$$

where

$$n! = \lfloor n = 1 \cdot 2 \cdot 3 \, (n - 1) \ldots\ldots\ldots n.$$

Example $5! = 5 \times 4 \times 3 \times 2 \times 1$

7 EXPANSIONS AND FACTORS

(i) $(a \pm b)^2 = a^2 \pm 2ab + b^2.$

(ii) $(a \pm b)^3 = a^3 \pm 3a^2b + 3ab^2 \pm b^3.$

(iii) $(a + b + c)^2 = a^2 + b^2 + c^2 + 2ab + 2ac + 2bc.$

(iv) $a^2 - b^2 = (a - b)(a + b).$

(v) $a^3 - b^3 = (a - b)(a^2 + ab + b^2).$

(vi) $a^3 + b^3 = (a + b)(a^2 - ab + b^2).$

(vii) $a^n - b^n = (a - b)(a^{n-1} + a^{n-2}b + \dots + b^{n-1}).$

(viii) $a^n - b^n = (a + b)(a^{n-1} - a^{n-2}b + \dots - b^{n-1})$, for n an even integer.

(ix) $a^n + b^n = (a + b)(a^{n-1} - a^{n-2}b + \dots + b^{n-1})$, for n an odd integer.

(x) $a^4 + a^2b^2 + b^4 = (a^2 + ab + b^2)(a^2 - ab + b^2).$

8 RATIO AND PROPORTION

If $a : b = c : d$ or $\dfrac{a}{b} = \dfrac{c}{d}$, then $ad = bc$, $\dfrac{a}{b} = \dfrac{c}{d}$.

If $\dfrac{a}{b} = \dfrac{c}{d} = \dfrac{e}{f} = \dots = k$, then

$$k = \frac{a + c + e + \dots}{b + d + f + \dots} = \frac{pa + qc + re + \dots}{pb + qd + rf + \dots}$$

9 CONSTANT FACTOR OF PROPORTIONALITY (OR VARIATION), k

(i) If y varies directly as x, or y is proportional to x,

$$y = kx.$$

(ii) If y varies inversely as x, or y is inversely proportional to x,

$$y = \frac{k}{x},$$

(iii) If y varies jointly as x and z,

$$y = kxz.$$

(viii) If y varies directly as x and inversely as z,

$$y = \frac{kx}{z}.$$

10 ARITHMETIC PROGRESSION

$$a, a + d, a + 2d, a + 3d, \ldots.$$

If a is the first term, d is the common difference, n is the number of terms, l is the last term, and S is the sum of n terms.

$$l = a + (n - 1)\,d,$$

the sum of the series upto l

$$S = \frac{n}{2}(a + l) = \frac{n}{2}[a + a + (n-1)d] = \frac{n}{2}[2a + (n-1)d]$$

The *arithmetic mean* of a and b is $(a + b)/2$.

11 GEOMETRIC PROGRESSION

$$a, ar, ar^2, ar^3, \ldots.$$

If a is the first term, r is the common ratio, n is the number of terms, l is the last term, and S_n is the sum of n terms,

$$l = ar^{n-1}, \qquad \text{sum up to } n \text{ terms } S_n = a\left(\frac{r^n - 1}{r - 1}\right) = \frac{rl - a}{r - 1}.$$

If $r^2 < 1$, S_n approaches the limit S_∞ as n increases without limit,

$$S_\infty = \frac{a}{1 - r}.$$

The *geometric mean* of a and b is $\sqrt{ab}$.

12   HARMONIC PROGRESSION

A sequence of numbers whose reciprocals form an arithmetic progression is called a *harmonic progression*. Thus

$$\frac{1}{a}, \frac{1}{a+d}, \frac{1}{a+2d}, \cdots$$

The *harmonic mean* of a and b is $2ab/(a + b)$.

13   PERMUTATIONS

Each different arrangement of all or a part of a set of things is called a *permutation*. The number of permutations of n different things taken r at a time is

$$P(n, r) = {_nP_r} = n(n-1)(n-2) \ldots (n-r+1) = \frac{n!}{(n-r)!},$$

where

$$n! = n(n-1)(n-2) \ldots (1).$$

14   COMBINATIONS

Each of the groups or relations which can be made by taking part or all of a set of things, without regard to the arrangement of the things in a group, is called a *combination*. The number of combinations of n different things taken r at a time is

$$C(n, r) = {^nC_r} = \binom{n}{r} = \frac{{^nP_r}}{r!} = \frac{n(n-1)\ldots(n-r+1)}{r(r-1)\ldots(1)} = \frac{n!}{r!(n-r)!}.$$

15   PROBABILITY

If an event may occur in p ways and may fail in q ways, all ways being equally likely, the *probability* of its occurrence is $p/(p + q)$, and that of its failure to occur is $q/(p + q)$.

16   REMAINDER THEOREM

If the polynomial $f(x)$ is divided by $(x - a)$, the remainder is $f(a)$. Hence, if a is a root of equation $f(x) = 0$, then $f(x)$ is divisible by $(x - a)$.

17   DETERMINANTS

The determinant D of order n,

$$D = \begin{vmatrix} a_{11} & a_{12} & \cdot & \cdot & \cdot & a_{1n} \\ a_{21} & a_{22} & \cdot & \cdot & \cdot & a_{2n} \\ \vdots & \vdots & \vdots & \vdots & \vdots & \vdots \\ a_{n1} & a_{n2} & \cdot & \cdot & \cdot & a_{nn} \end{vmatrix}$$

is defined to be the sum

$$\Sigma (\pm)\, a_{1i} a_{2j} a_{3k} \ldots a_{nl}$$

of $n!$ terms, the sign in a given term being taken plus or minus according to the number of inversions (of the numbers $1, 2, 3, \ldots, n$) in the corresponding sequence $i, j, k, \ldots, l$, is even or odd.

The *cofactor* A_{ij} of the element a_{ij} is defined to be the product of $(-1)^{i+j}$ by the determinant obtained from D by deleting the ith row and the jth column.

The following theorems are true:

(a) If the corresponding rows and columns of D be interchanged, D is unchanged.

(b) If any two rows (or columns) of D be interchanged, D is changed to $-D$.

(c) If any two rows (or columns) are alike, then $D = 0$.

(d) If each element of a row (or column) of D be multiplied by m, the new determinant is equal to mD.

(e) If each element of a row (or column) is added m times the corresponding element in another row (or column), D is unchanged.

(f) $D = a_{1j}A_{1j} + a_{2j}A_{2j} + \ldots + a_{nj}A_{nj},$ $\qquad j = 1, 2, \ldots\ldots, n.$

(g) $0 = a_{1k}A_{1j} + a_{2k}A_{2j} + \ldots\ldots + a_{nk}A_{nj},$ $\quad$ if $j \neq k.$

(h) The solution of the system of equations

$$a_{i1}x_1 + a_{i2}x_2 + \ldots + a_{in}x_n = c_i, \ i = 1, 2, \ldots, n,$$

is unique if $D \neq 0$. The solution is given by the equations

$$Dx_1 = C_1, \ Dx_2 = C_2, \ \ldots, \ Dx_n = C_n,$$

where C_k is what D becomes when the elements of its kth column are replaced by $c_1, c_2, \ldots, c_n$, respectively.

EXAMPLE 1.

$$\begin{vmatrix} a & b & c \\ p & q & r \\ l & m & n \end{vmatrix}$$

$$= a\begin{vmatrix} q & r \\ m & n \end{vmatrix} - b\begin{vmatrix} p & r \\ l & n \end{vmatrix} + c\begin{vmatrix} p & q \\ l & m \end{vmatrix}$$

$$= a(qn - rm) - b\,(pn - rl) + c\,(pm - ql).$$

EXAMPLE 2.

Find the values of x, y, and z, which satisfy the system

$$3x + 2y + z = 1000$$

$$4x + y + 3z = 1500$$

$$x + y + z = 600$$

Solution:

Here,

$$x = \frac{\begin{vmatrix} 1000 & 2 & 1 \\ 1500 & 1 & 3 \\ 600 & 1 & 1 \end{vmatrix}}{\begin{vmatrix} 3 & 2 & 1 \\ 4 & 1 & 3 \\ 1 & 1 & 1 \end{vmatrix}} = \frac{-500}{-5} = 100$$

$$y = \frac{\begin{vmatrix} 3 & 1000 & 1 \\ 4 & 1500 & 3 \\ 1 & 600 & 1 \end{vmatrix}}{\begin{vmatrix} 3 & 2 & 1 \\ 4 & 1 & 3 \\ 1 & 1 & 1 \end{vmatrix}} = \frac{-1000}{-5} = 200$$

$$z = \frac{\begin{vmatrix} 3 & 2 & 1000 \\ 4 & 1 & 1500 \\ 1 & 1 & 600 \end{vmatrix}}{\begin{vmatrix} 3 & 2 & 1 \\ 4 & 1 & 3 \\ 1 & 1 & 1 \end{vmatrix}} = \frac{-1500}{-5} = 300$$

Interest, Annuities, Sinking Funds.

In this section, n is the number of years, and r is the rate of interest expressed as a decimal.

18   AMOUNT

A principal P placed at a rate of interest r for n years accumulates to an amount A, as follows:

At simple interest: $\qquad\qquad\qquad\qquad A = P\,(1 + nr).$

At interest compounded annually: $\qquad A = P(1 + r)^n.$

At interest compounded q times a year: $A = P\left(1 + \frac{4}{q}\right)^{nq}.$

19   NOMINAL AND EFFECTIVE RATES

The rate of interest quoted in describing a given variety of compound interest is called the *nominal rate*. The rate per year at which interest is earned during each year is called the *effective rate*. The effective rate i corresponding to the nominal rate r, compounded q times a year is:

$$i = \left(1 + \frac{r}{q}\right)^q - 1.$$

20   PRESENT OR DISCOUNTED VALUE OF A FUTURE AMOUNT

The present quantity P which in n years will accumulate to the amount A at the rate of interest r, is:

At simple interest:
$$P = \frac{A}{1 + nr}.$$

At interest compounded annually:
$$P = \frac{A}{(1 + r)^n}.$$

At interest compounded q times a year:
$$P = \frac{A}{\left(1 + \dfrac{r}{q}\right)^{nq}}.$$

P is called the *present value* of A due in n years at rate r.

21   TRUE DISCOUNT

The true discount is:

$$D = A - P.$$

22   ANNUITY

A fixed sum of money paid at regular intervals is called an *annuity*.

23   AMOUNT OF AN ANNUITY

If an annuity P is deposited at the end of each successive year (beginning one year hence), and the interest at rate r, compounded annually, is paid on the accumulated deposit at the end of each year, the total amount N accumulated at the end of n years is

$$N = P.\frac{(1 + r)^n - 1}{r}$$

N is called the *amount* of an annuity P.

24   PRESENT VALUE OF AN ANNUITY

The total present amount P which will supply an annuity N at the end of each year for n years, beginning one year hence (assuming that in successive years the amount not yet paid out earns interest at rate r, compounded annually), is:

$$P = N \cdot \frac{(1+r)^n - 1}{r(1+r)^n} = N \cdot \frac{1 - (1+r)^{-n}}{r}.$$

P is called the *present value* of an annuity.

25   AMOUNT OF A SINKING FUND

If a fixed investment N is made at the end of each successive year (beginning at the end of the first year), and interest paid at rate r, compounded annually, is paid on the accumulated amount of the investment at the end of each year, the total amount S accumulated at the end of n years is:

$$S = N \cdot \frac{(1+r)^n - 1}{r}.$$

S is called the *amount of the sinking fund*.

26   FIXED INVESTMENT, OR ANNUAL INSTALLMENT

The amount N that must be placed at the end of each year (beginning one year hence), with compound interest paid at rate r on the accumulated deposit, in order to accumulate a sinking fund S in n years is:

$$N = S \cdot \frac{r}{(1+r)^n - 1}.$$

N is called a *fixed investment* or *annual installment*.

Algebraic Equations

27   QUADRATIC EQUATIONS

If $ax^2 + bx + c = 0,\ a \neq 0,$

then, $x = \dfrac{-b \pm \sqrt{b^2 - 4ac}}{2a}.$

If $a, b,$ and c are real and

if $b^2 - 4ac > 0$, the roots are real and unequal;

if $b^2 - 4ac = 0$, the roots are real and equal, and;

if $b^2 - 4ac < 0$, the roots are imaginary.

28   CUBIC EQUATIONS

The cubic equation

$$y^3 + py^2 + qy + r = 0,$$

may be reduced by the substitution

$$y = \left(x - \frac{p}{3} \right)$$

to the normal form

$$x^3 + ax + b = 0$$

where

$$a = \frac{1}{3}\left(3q - p^2\right),\ b = \frac{1}{27}\left(2p^3 - 9pq + 27r\right),$$

which has the solutions $x_1,\ x_2,$ and $x_3,$

$$x_1 = A + B,\ x_2,\ x_3 = -\frac{1}{2}(A + B) \pm \frac{i\sqrt{3}}{2}(A - B),$$

where

$$i^2 = -1, \; A = \sqrt[3]{-\frac{b}{2} + \sqrt{\frac{b^2}{4} + \frac{a^3}{27}}} \quad B = \sqrt[3]{-\frac{b}{2} - \sqrt{\frac{b^2}{4} + \frac{a^3}{27}}}.$$

If p, q, and r are real (and hence, if a and b are real), and

if $\dfrac{b^2}{4} + \dfrac{a^3}{27} > 0,$ there are one real root and two conjugate imag-

inary roots,

if $\dfrac{b^2}{4} + \dfrac{a^3}{27} = 0,$ there are three real roots of which at least two

are equal,

if $\dfrac{b^2}{4} + \dfrac{a^3}{27} < 0,$ there are three real and unequal roots.

If $\dfrac{b^2}{4} + \dfrac{a^3}{27} < 0.$ the above formulas are impractical. The real

roots are,

$$x_k = 2\sqrt{-\frac{a}{3}} \cos\left(\frac{\phi}{3} + 120°k\right), \qquad k = 0, 1, 2,$$

where

$$\cos \phi = \mp \sqrt{\frac{b^2}{4} \div \left(-\frac{a^3}{27}\right)},$$

and where the upper sign is to be used if b is positive and the lower sign if b is negative.

If $\qquad \dfrac{b^2}{4} + \dfrac{a^3}{27} > 0, \quad$ and $a > 0$

then the real root is,

$$x = 2\sqrt{\frac{a}{3}} \cot 2\, \phi,$$

where ϕ and Ψ are to be computed from

$$\cot 2\,\psi = \mp\sqrt{\dfrac{b^2}{4} \div \dfrac{a^3}{27}}\,, \quad \tan\phi = \sqrt[3]{\tan\psi}$$

and where the upper sign is to be used if b is positive and the lower sign if b is negative.

If
$$\dfrac{b^2}{4} + \dfrac{a^3}{27} = 0$$

the roots are,

$$x = \mp 2\sqrt{-\dfrac{a}{3}},\ \pm\sqrt{-\dfrac{a}{3}},\ \pm\sqrt{-\dfrac{a}{3}},$$

where the upper sign is to be used if b is positive, and the lower sign if b is negative.

29   BIQUADRATIC (QUARTIC) EQUATION

The quartic equation

$$y^4 + py^3 + qy^2 + ry + s = 0$$

may be reduced to the form

$$x^4 + ax^2 + bx + c = 0$$

by the substitution

$$y = \left(x - \dfrac{p}{4} \right).$$

Let l, m, and n denote the roots of the resolvent cubic.

$$t^3 + \left(\dfrac{a}{2}\right)t^2 + \left(\dfrac{a^2 - 4c}{16}\right)t - \dfrac{b^2}{64} = 0.$$

The required roots of the reduced quartic are,

$$
\begin{aligned}
x_1 &= +\sqrt{l} + \sqrt{m} + \sqrt{n}, & x_2 &= +\sqrt{l} - \sqrt{m} - \sqrt{n}, \\
x_3 &= -\sqrt{l} + \sqrt{m} - \sqrt{n}, & x_4 &= -\sqrt{l} - \sqrt{m} + \sqrt{n},
\end{aligned}
$$

where the selection of the square root to be attached to each of the quantities $\sqrt{l}$, $\sqrt{m}$, $\sqrt{n}$ must give $\sqrt{l}\,\sqrt{m}\,\sqrt{n} = -\dfrac{b}{8}$.

30   GENERAL EQUATIONS OF THE nth DEGREE

$$P \equiv a_0 x^n + a_1 x^{n-1} + a_2 x^{n-2} + \ldots + a_{n-1} x + a_n = 0.$$

If $n > 4$, there is no formula that gives the roots of this general equation. The following methods may be used to advantage:

(a) *Roots by Factors:* By trial, find a number r such that $x = 1$ satisfies the equation, that is, such that

$$a_0 r^n + a_1 r^{n-1} + a_2 r^{n-2} + \ldots + a_{n-1} r + a_n = 0, \, a_0 \neq 0.$$

Then $(x - r)$ is a factor of the left-hand member P of the equation. Divide P by $(x - r)$ leaving an equation of degree one less than that of the original equation. Next, proceed in the same manner with the reduced equation. (All integer roots of $P = 0$ are divisors of a_n/a_0.)

(b) *Roots by Approximation:* Suppose the coefficients a_i in P are real. Let a and b be real numbers. If for $x = a$ and $x = b$, the left member P of the equation has opposite signs, then a root lies between a and b. By repeated application of this principle, real roots to any desired degree of accuracy may be obtained.

(c) *Roots by Graphing:* If a graph of P is plotted as a function of x, the real roots are the values of x where the graph crosses the x-axis. By increasing the scale of the portion of the graph near an estimated root, the root may be obtained to any desired degree of accuracy.

(d) *Descartes' Rule:* The number of positive real roots of an equation with real coefficients is either equal to the number of its variations of sign or is less than that number by a positive even integer. A root of multiplicity m is here counted as m roots.

31(a) THE EQUATION $x^n = a$.

The n roots of this equation are:

$$x = \sqrt[n]{a}\left(\cos\frac{2k\pi}{n} + \sqrt{-1}\,\sin\frac{2k\pi}{n}\right), \text{if } a > 0,$$

$$x = \sqrt[n]{-a}\left(\cos\frac{(2k+1)\pi}{n} + \sqrt{-1}\,\sin\frac{(2k+1)\pi}{n}\right), \text{if } a < 0,$$

where k takes successively the values 0, 1, 2, 3, ..., $n - 1$.

31(b) STIRLING'S FORMULA

For large values of n,

$$\sqrt{2n\pi}(n/e)^n < n! < \sqrt{2n\pi}(n/e)^n[1 + 1/12n - 1],$$

where π = 3.14159..., e = 2.71828...

$$\log n! \cong \left(n + \frac{1}{2}\right)\log n - n\log e + \log\sqrt{2\pi}.$$

ELEMENTARY GEOMETRY

Let a, b, c, d, and s denote lengths, A denote areas, and V denote volumes.

32   TRIANGLE

$A = bh/2$, where b denotes the base and h denotes the altitude.

33   RECTANGLE

$A = ab$, where a and b denote the lengths of the sides.

34   PARALLELOGRAM (OPPOSITE SIDES PARALLEL)

$A = ah = ab \sin \theta$, where a and b denote the sides, h denotes the altitude, and θ denotes the angle between the sides.

35   TRAPEZOID (FOUR SIDES, TWO PARALLEL)

$A = \dfrac{1}{2} h (a + b)$, where a and b are the sides and h the altitude.

36 REGULAR POLYGON OF *N* SIDES

$$A = \frac{1}{4} n \, a^2 \cot \frac{180°}{n},$$ where a is the length of side.

$$R = \frac{a}{2} \operatorname{cosec} \frac{180°}{n},$$ where R is the radius of circumscribed circle.

$$r = \frac{a}{2} \cot \frac{180°}{n},$$ where r is the radius of inscribed circle.

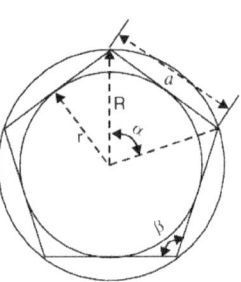

FIGURE 2.1.

$$\alpha = \frac{360°}{n} = \frac{2\pi}{n},$$ radians,

$$\beta = \left(\frac{n-2}{n}\right) \cdot 180° = \left(\frac{n-2}{n}\right) \pi$$ radians where α and β are the angles indicated in Figure 2.1

$$a = 2r \tan \frac{a}{2} = 2R \sin \frac{a}{2}.$$

37 CIRCLE

Let C = circumference, S = length of arc subtended by θ,

R = radius, l = chord subtended by arc S,

D = diameter, h = rise,

`A = area, θ = central angle in radians.

$C = 2\pi R = \pi D,$ $\pi = 3.14159...$

$$S = R\theta = \frac{1}{2} D$$ $$\theta = D \cos{-1} \frac{d}{R},$$

$$l = 2\sqrt{R^2 - d^2} = 2R \sin \frac{\theta}{2} = 2d \tan \frac{\theta}{2}.$$

$$d = \frac{1}{2}\sqrt{4R^2 - l^2} = R \cos \frac{\theta}{2} = \frac{1}{2} l \operatorname{ctn} \frac{\theta}{2}.$$

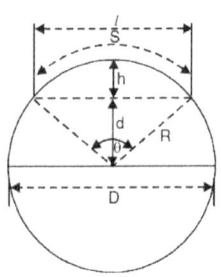

FIGURE 2.2.

$h = R - d.$

$$\theta = \frac{S}{R} = \frac{2S}{D} = 2\cos^{-1}\frac{d}{R} = 2\tan^{-1}\frac{1}{2d} = 2\sin^{-1}\frac{l}{D}.$$

$$A\ (\text{circle}) = \pi R2 = \frac{1}{4}\pi D2.$$

$$A\ (\text{sector}) = \frac{1}{2}Rs = \frac{1}{2}R^2\theta.$$

$$A\ (\text{segment}) = A\ (\text{sector}) - A\ (\text{triangle}) = \frac{1}{2}R^2(\theta - \sin\theta)$$

$$= R^2\cos^{-1}\frac{(R-h)}{R} - (R-h)\sqrt{2Rh - h^2}.$$

Perimeter of an n-side regular polygon inscribed *in a circle.*

$$= 2n\ R\ \sin n\ \frac{\pi}{n}.$$

Area of inscribed polygon $= \dfrac{1}{2}\ nR^2\ \sin\dfrac{2\pi}{n}.$

Perimeter of an n-side regular polygon circumscribed *about a circle*

$$= 2n\ R\ \tan\frac{\pi}{n}.$$

Area of circumscribed polygon $= nR^2\ \tan\dfrac{\pi}{n}.$

Radius of a circle inscribed in a triangle of sides a, b, and c is

$$r = \sqrt{\frac{(s-a)(s-b)(s-c)}{s}}, \quad s = \frac{1}{2}(a+b+c).$$

Radius of a circle circumscribed about a triangle is

$$R = \frac{abc}{4\sqrt{s(s-a)(s-b)(s-c)}}.$$

38   ELLIPSE

Perimeter $= 4a.\ E(k)$, where $k^2 = 1 - (b^2/a^2)$ and $E(k)$ is the complete elliptic integral $E.$

$A = \pi ab$, where a and b are lengths of semi-major and semi-minor axes, respectively.

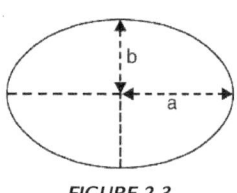

FIGURE 2.3.

39   PARABOLA

$$A = 2ld/3.$$

$$\text{Height of } d_1 = \frac{d}{l^2}\left(l^2 - l_1^2\right).$$

$$\text{Width of } l_1 = l\sqrt{\frac{d - d_1}{d}}.$$

$$\text{Length of arc} = l\left[1 + \frac{2}{3}\left(\frac{2d}{l}\right)^2 - \frac{2}{5}\left(\frac{2d}{l}\right)^4 + \ldots\right].$$

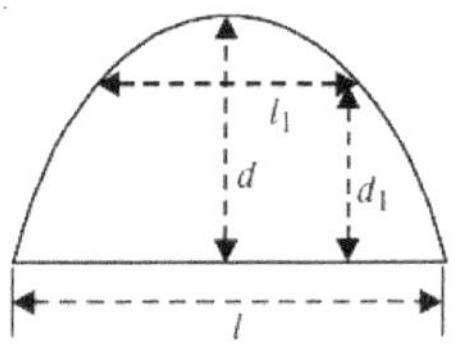

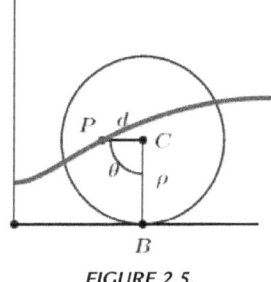

FIGURE 2.4.

40   CATENARY, CYCLOID, ETC.

Let P be a point at a distance d from the center of a circle of radius ρ. The curve traced out P as the circle rolls along a straight line, without slipping, is called a cycloid, as shown in Figure 2.5.

FIGURE 2.5.

41   AREA BY APPROXIMATION

If $y_0, y_1, y_2, \ldots, y_n$ be the lengths of a series of equally spaced parallel chords and if h is their distance apart, the area enclosed by boundary is given approximately by any one of the following formulae:

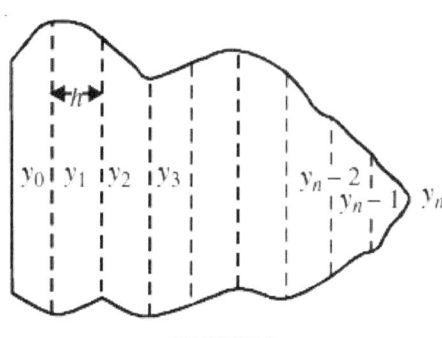

FIGURE 2.6.

$$A_T = h\left[\frac{1}{2}(y_0 + y_n) + y_1 + y_2 + \ldots + y_{n-1}\right]. \text{ (Trapezoidal Rule)}.$$

$$A_D = h\left[0.4(y_0 + y_n) + 1.1(y_1 + y_{n-1}) + y_2 + y_3 + \ldots + y_{n-2}\right].$$
$$\text{(Durand's Rule)}.$$

$$A_S = \frac{1}{3}h\left[(y_0 + y_n) + 4(y_1 + y_3 + \ldots + y_{n-1}) + 2(y_2 + y_4 + \ldots + y_{n-2})\right].$$
$$(n \text{ even, Simpson's Rule)}.$$

In general, *As* gives a more accurate approximation.

The greater the value of n, the greater the accuracy of approximation.

42   CUBE

$V = a^3$; $d = a\sqrt{3}$; total surface area $= 6a^2$, where a is the length of side and d is the length of diagonal.

43   RECTANGULAR PARALLELOPIPED

$V = abc$; $d = \sqrt{a^2 + b^2 + c^2}$; total surface area $= 2(ab + bc + ca)$, where a, b and c are the lengths of the sides and d is the length of diagonal.

44   PRISM OR CYLINDER

$$V = (\text{area of base}) \cdot (\text{altitude}).$$

Lateral area = (perimeter of right section). (lateral edge).

45   PYRAMID OR CONE

$$V = \frac{1}{3}(\text{area of base}) \cdot (\text{altitude})$$

Lateral area of regular pyramid $= \frac{1}{2}(\text{perimeter of base}) \cdot (\text{slant height})$.

46 (A)   FRUSTUM OF PYRAMID OR CONE

$V = \dfrac{1}{3}\left(A_1 + A_2 + \sqrt{A_1 \cdot A_2}\right)h$, where h is the altitude and $A1$ and $A2$ are the areas of the bases.

Lateral area of a regular figure $= \dfrac{1}{2}$ (sum of perimeters of base) $\cdot$ (slant height).

46 (B)   PRISMOID

$V = \dfrac{h}{6}\,(A_1 + A_2 + 4A_3)$, where $h =$ altitude, A_1 and A_2 are the areas

of the bases and A_3 is the area of the midsection parallel to the bases.

47   AREA OF SURFACE AND VOLUME OF REGULAR POLYHEDRA OF EDGE *L*

Name	Type of Surface	Area of Surface	Volume
Tetrahedron	4 equilateral triangles	$1.73205l_2$	$0.11785l_3$
Hexahedron (cube)	6 squares	$6.00000l_2$	$1.00000l_3$
Octahedron	8 equilateral triangles	$3.46410l_2$	$0.47140l_3$
Dodecahedron	12 pentagons	$20.64578l_2$	$7.66312l_3$
Icosahedron	20 equilateral triangles	$8.66025l_2$	$2.18169l_3$

48   SPHERE

A (sphere) $= 4\,\pi R^2 = \pi D^2$.

A (zone) $= 2\,\pi Rh_1 = \pi Dh_1$.

V (sphere) $= \dfrac{4}{3}\pi R^3 = \dfrac{1}{6}\pi D^3$.

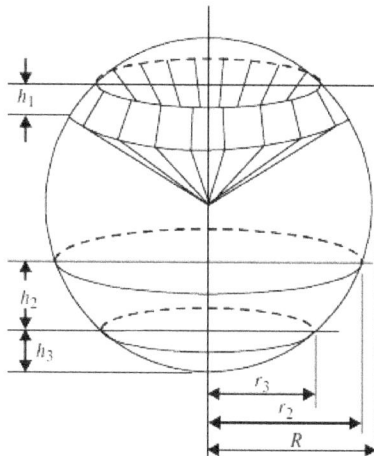

FIGURE 2.7.

$$V \text{ (spherical sector)} = \frac{2}{3}\pi R^2 \; h_1 = \frac{1}{6}\pi D^2 h_1 .$$

$$V \text{ (spherical segment of one base)} = \frac{1}{6}\pi h_3 \left(3r_3^2 + h_3^2\right).$$

$$V \text{ (spherical segment of two bases)} = \frac{1}{6}\pi h_2 \left(3r_3^2 + 3r_2^2 + h_2^2\right).$$

A (lune) $= 2R_2\theta$, where θ is the angle in radians of lune.

49   SOLID ANGLE (Ψ)

The solid angle ψ at any point subtended by a surface S is equal to the area A of the portion of the surface of a sphere of unit radius, center at P, which is cut out by a conical surface, with vertex at P, passing through the perimeter of S.

The unit solid angle ψ is called the *steradian*.

The total solid angle about a point is 4π *steradians*.

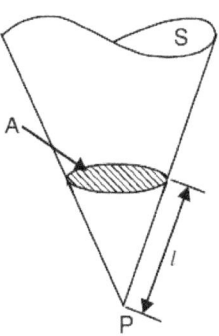

FIGURE 2.8.

50   ELLIPSOID

$V = \dfrac{4}{3}\pi abc$, where a, b, and c are the lengths of the semi-axes.

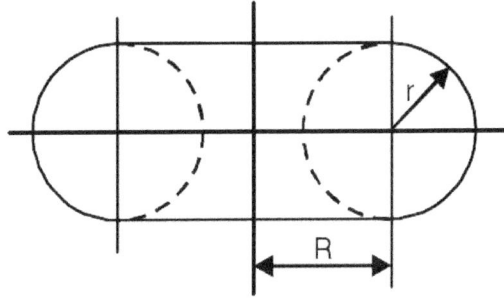

FIGURE 2.9.

51 (*A*)   TORUS

$$V = 2\pi 2 R r 2.$$
$$\text{Area of surface } = S = 4\pi 2 R r.$$

51 (*B*)   THEOREMS OF PAPPUS

(a) If a plane area A is rotated about a line l in the plane of A and not cutting A, the volume of the solid generated is equal to the product of A and the distance traveled by the center of gravity of A.

(b) If a plane curve C is rotated about a line l in the plane of C and not cutting C, the area of the surface generated is equal to the product of the length of C and the distance traveled by the center of gravity of C.

3

*T*RIGONOMETRY

52   ANGLE

If two lines intersect, one line may be rotated about their point of intersection through the *angle* which they form until it coincides with the other line.

The angle is said to be *positive* if the rotation is counterclockwise and *negative,* if clockwise.

A complete revolution of a line is a rotation through an angle of 360°. Thus,

A *degree* is 1/360 of the plane angle about a point.

A *radian* is an angle subtended at the center of a circle by an arc whose length is equal to that of the radius.

$$180° = \pi \text{ radians}; \quad 1° = \frac{\pi}{180} \text{ radians}; \quad 1 \text{ rad.} = \frac{180}{\pi} \text{ degrees.}$$

53   TRIGONOMETRIC FUNCTIONS OF AN ANGLE α

Let α be any angle whose initial sidelines on the positive x-axis and whose vertex is at the origin, and (x, y) be any point on the terminal side of the angle. (x is positive if measured along OX to the right, from the y-axis; and negative, if measured along OX' to the left from the y-axis. Likewise, y is positive if measured parallel to OY, and negative if measured parallel to OY'.) Let r be the positive

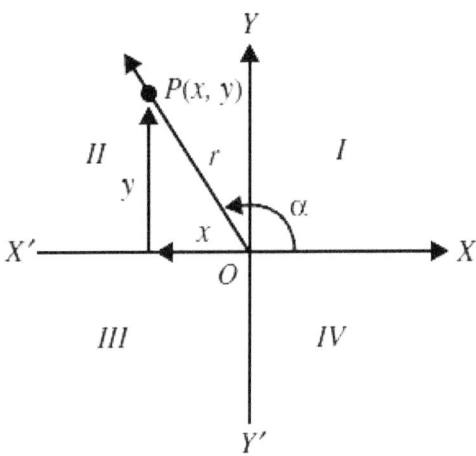

FIGURE 3.1.

distance from the origin to the point. The trigonometric functions of an angle are defined as follows:

sine $\alpha = \sin \alpha = \dfrac{y}{r}$.

cosine $\alpha = \cos \alpha = \dfrac{x}{r}$.

tangent $\alpha = \tan \alpha = \dfrac{y}{x}$.

cotangent $\alpha = \text{ctn } \alpha = \cot \alpha = \dfrac{x}{y}$.

secant $\alpha = \sec \alpha = \dfrac{r}{x}$.

cosecant $\alpha = \csc \alpha = \text{cosec } \alpha = \dfrac{r}{y}$.

exsecant $\alpha = \text{exsec } \alpha = \sec \alpha - 1$.

versine $\alpha = \text{vers } \alpha = 1 - \cos \alpha$.

coversine $\alpha = \text{covers } \alpha = 1 - \sin \alpha$.

haversine $\alpha = \text{hav } \alpha = \dfrac{1}{2} \text{ vers } \alpha$

54 SIGNS OF THE FUNCTIONS

Quadrant	sin	cos	tan	ctn	sec	csc
I..................	+	+	+	+	+	+
II.................	+	−	−	−	−	+
III...............	−	−	+	+	−	−
IV................	−	+	−	−	+	−

55 FUNCTIONS OF 0°, 30°, 45°, 60°, 90°, 180°, 270°, 360°

	0°	30°	45°	60°	90°	180°	270°	360°
sin	0	$\dfrac{1}{2}$	$\dfrac{1}{\sqrt{2}}$	$\dfrac{\sqrt{3}}{2}$	1	0	−1	0
cos	1	$\dfrac{\sqrt{3}}{2}$	$\dfrac{1}{\sqrt{2}}$	$\dfrac{1}{2}$	0	−1	0	1
tan	0	$\dfrac{1}{\sqrt{3}}$	1	$\sqrt{3}$	∞	0	∞	0
cot	∞	$\sqrt{3}$	1	$\dfrac{1}{\sqrt{3}}$	0	∞	0	∞
sec	1	$\dfrac{2}{\sqrt{3}}$	$\sqrt{2}$	2	∞	−1	∞	1
cosec	∞	2	$\sqrt{2}$	$\dfrac{2}{\sqrt{3}}$	1	∞	−1	∞

56 VARIATIONS OF THE FUNCTIONS

Quadrant	sin	cos	tan	ctn	sec	csc
I	$0 \to +1$	$+1 \to 0$	$0 \to +\infty$	$+\infty \to 0$	$+1 \to +\infty$	$+\infty \to +1$
II	$+1 \to 0$	$0 \to -1$	$-\infty \to 0$	$0 \to -\infty$	$-\infty \to -1$	$+1 \to +\infty$
III	$0 \to -1$	$-1 \to 0$	$0 \to +\infty$	$+\infty \to 0$	$-1 \to -\infty$	$-\infty \to -1$
IV	$-1 \to 0$	$0 \to +1$	$-\infty \to 0$	$0 \to -\infty$	$+\infty \to +1$	$-1 \to -\infty$

57 FUNCTIONS OF ANGLES IN ANY QUADRANT IN TERMS OF ANGLES IN FIRST QUADRANT

	$-\alpha$	$90° \pm \alpha$	$180° \pm \alpha$	$270° \pm \alpha$	$n(360)°$ $\pm \alpha$
sin	$-\sin \alpha$	$+\cos \alpha$	$\mp \sin \alpha$	$-\cos \alpha$	$\pm \sin \alpha$
cos	$+\cos \alpha$	$\mp \sin \alpha$	$-\cos \alpha$	$\pm\sin \alpha$	$+\cos \alpha$
tan	$-\tan \alpha$	$\mp \operatorname{ctn} \alpha$	$\pm \tan \alpha$	$\mp \operatorname{ctn} \alpha$	$\pm \tan \alpha$
cot	$-\operatorname{ctn} \alpha$	$\mp \tan \alpha$	$\pm \operatorname{ctn} \alpha$	$\mp \tan \alpha$	$\pm \operatorname{ctn} \alpha$
sec	$+\sec \alpha$	$\mp \csc \alpha$	$-\sec \alpha$	$\pm \csc \alpha$	$+\sec \alpha$
cosec	$-\csc \alpha$	$+\sec \alpha$	$\mp \csc \alpha$	$-\sec \alpha$	$\pm \csc \alpha$

where n denotes any integer.

58 FUNDAMENTAL IDENTITIES

$$\sin \alpha = \frac{1}{\csc \alpha}; \cos \alpha = \frac{1}{\sec \alpha}; \tan \alpha = \frac{1}{\cot \alpha} = \frac{\sin \alpha}{\cos \alpha}.$$

$$\operatorname{cosec} \alpha = \frac{1}{\sin \alpha}; \sec \alpha = \frac{1}{\cos \alpha}; \cot \alpha = \frac{1}{\tan \alpha} = \frac{\cos \alpha}{\sin \alpha}.$$

$$\sin^2 \alpha + \cos^2 \alpha = 1; 1 + \tan^2 \alpha = \sec^2 \alpha; 1 + \cot^2\alpha = \operatorname{cosec}^2 \alpha.$$

$$\sin 2\alpha = 2 \sin \alpha \cos \alpha$$

$$\cos 2\alpha = 2 \cos^2 \alpha - 1 = 1 - 2 \sin^2 \alpha = \cos^2 \alpha - \sin^2 - \alpha,$$

$$\tan 2\alpha = \frac{2\tan \alpha}{1 - \tan^2 \alpha},$$

$$\sin 3\alpha = 3 \sin \alpha - 4 \sin^3 \alpha,$$

$$\cos 3\alpha = 4 \cos^3 \alpha - 3 \cos \alpha,$$

$$\sin n\alpha = 2 \sin (n-1)\alpha \cdot \cos \alpha - \sin (n-2)\alpha,$$

$$\cos n\alpha = 2 \cos (n-1)\alpha \cdot \cos \alpha - \cos (n-2)\alpha,$$

$$\sin (\alpha \pm \beta) = \sin \alpha \cos \beta \pm \cos \alpha \sin \beta,$$

$$\cos (\alpha \pm \beta) = \cos \alpha \cos\beta \mp \sin \alpha \sin \beta,$$

$$\tan (\alpha \pm \beta) = \frac{\tan \alpha \pm \tan \beta}{1 \mp \tan \alpha \tan \beta}.$$

$$\sin \alpha + \sin \beta = 2 \sin \frac{1}{2}(\alpha + \beta) \cdot \cos \frac{1}{2}(\alpha - \beta).$$

$$\sin \alpha - \sin \beta = 2 \cos \frac{1}{2}(\alpha + \beta) \cdot \sin \frac{1}{2}(\alpha - \beta),$$

$$\cos \alpha + \cos \beta = 2 \cos \frac{1}{2}(\alpha + \beta) \cdot \cos \frac{1}{2}(\alpha - \beta),$$

$$\cos \alpha - \cos \beta = -2 \sin \frac{1}{2}(\alpha + \beta) \cdot \sin \frac{1}{2}(\alpha - \beta).$$

$$\sin \frac{\alpha}{2} = \pm \sqrt{\frac{1 - \cos \alpha}{2}}, \text{ positive if } \frac{\alpha}{2} \text{ in I or II quadrants,}$$
negative otherwise.

$$\cos \frac{\alpha}{2} = \pm \sqrt{\frac{1 + \cos \alpha}{2}}, \text{ positive if } \frac{\alpha}{2} \text{ in I or IV, negative}$$
otherwise.

$$\frac{\tan \alpha}{2} = \frac{1 - \cos \alpha}{\sin \alpha} = \frac{\sin \alpha}{1 + \cos \alpha} = \pm \sqrt{\frac{1 - \cos \alpha}{1 + \cos \alpha}},$$

positive if $\frac{\alpha}{2}$ in I or III, negative otherwise.

$$\sin^2 \alpha = \frac{1}{2}(1 - \cos 2\alpha); \cos^2 \alpha = \frac{1}{2}(1 + \cos 2\alpha)$$

$$\sin^3 \alpha = \frac{1}{4}(3 \sin \alpha - \sin 3\alpha); \cos^3 \alpha = \frac{1}{4}(\cos 3\alpha + 3 \cos \alpha).$$

$$\sin \alpha \sin \beta = \frac{1}{2} \cos (\alpha - \beta) - \frac{1}{2} \cos (\alpha + \beta).$$

$$\cos \alpha \cos \beta = \frac{1}{2} \cos (\alpha - \beta) + \frac{1}{2} \cos (\alpha + \beta),$$

$$\sin \alpha \sin \beta = \frac{1}{2} \sin (\alpha + \beta) + \frac{1}{2} \sin (\alpha - \beta).$$

59 EQUIVALENT EXPRESSIONS FOR SIN, COS, TAN, ETC.

Function	sin α	cos α	tan α	cot α	sec α	cosec α
sin α	$\sin \alpha$	$\pm\sqrt{1-\cos^2 \alpha}$	$\dfrac{\tan \alpha}{\pm\sqrt{1+\tan^2 \alpha}}$	$\dfrac{1}{\pm\sqrt{1+\cot^2 \alpha}}$	$\dfrac{\pm\sqrt{\sec^2 \alpha-1}}{\sec \alpha}$	$\dfrac{1}{\operatorname{cosec} \alpha}$
cos α	$\pm\sqrt{1-\sin^2 \alpha}$	$\cos \alpha$	$\dfrac{1}{\pm\sqrt{1+\tan^2 \alpha}}$	$\dfrac{\cot \alpha}{\pm\sqrt{1+\cot^2 \alpha}}$	$\dfrac{1}{\sec \alpha}$	$\dfrac{\pm\sqrt{\operatorname{cosec}^2 \alpha-1}}{\operatorname{cosec} \alpha}$
tan α	$\dfrac{\sin \alpha}{\pm\sqrt{1-\sin^2 \alpha}}$	$\dfrac{\pm\sqrt{1-\cos^2 \alpha}}{\cos \alpha}$	$\tan \alpha$	$\dfrac{1}{\cot \alpha}$	$\pm\sqrt{\sec^2 \alpha-1}$	$\dfrac{1}{\pm\sqrt{\operatorname{cosec}^2 \alpha-1}}$
cot α	$\dfrac{\pm\sqrt{1-\sin^2 \alpha}}{\sin \alpha}$	$\dfrac{\cos \alpha}{\pm\sqrt{1-\cos^2 \alpha}}$	$\dfrac{1}{\tan \alpha}$	$\cot \alpha$	$\dfrac{1}{\pm\sqrt{\sec^2 \alpha-1}}$	$\pm\sqrt{\operatorname{cosec}^2 \alpha-1}$
sec α	$\dfrac{1}{\pm\sqrt{1-\sin^2 \alpha}}$	$\dfrac{1}{\cos \alpha}$	$\pm\sqrt{1+\tan^2 \alpha}$	$\dfrac{\pm\sqrt{1+\cot^2 \alpha}}{\cot \alpha}$	$\sec \alpha$	$\dfrac{\operatorname{cosec} \alpha}{\pm\sqrt{\operatorname{cosec}^2 \alpha-1}}$
cosec α	$\dfrac{1}{\sin \alpha}$	$\dfrac{1}{\pm\sqrt{1-\cos^2 \alpha}}$	$\dfrac{\pm\sqrt{1+\tan^2 \alpha}}{\tan \alpha}$	$\pm\sqrt{1+\cot^2 \alpha}$	$\dfrac{\sec \alpha}{\pm\sqrt{\sec^2 \alpha-1}}$	$\operatorname{cosec} \alpha$

Note: The quadrant which terminates determines the sign used.

60 INVERSE OR ANTI-TRIGONOMETRIC FUNCTIONS

The complete solution of the equation $x = \sin y$ is (in radians):

$$y = (-1)^n \sin^{-1} x + n(\pi), \qquad -\pi/2 \le \sin^{-1} \le x/2,$$

where $\sin^{-1} x$ is the *principal value* of the angle whose sine is x (*n* denotes an integer).

Similarly, if $x = \cos y$,

$$y = \pm \cos^{-1} x + n(2\pi), \; 0 \le \cos^{-1} x \le \pi.$$

If $x = \tan y$,

$$y = \tan^{-1} x + n(\pi), \; -\pi/2 < \tan^{-1} x < \pi/2.$$

Similar relations hold for $x = \cot y$, $x = \sec y$, and $x = \operatorname{cosec} y$.

61   CERTAIN RELATIONS AMONG INVERSE FUNCTIONS

If the inverse functions are restricted as in Art. 60, the following formulae hold:

$$\sin^{-1} a = \cos^{-1} \sqrt{1-a^2} = \tan^{-1} \frac{a}{\sqrt{1-a^2}} = \cot^{-1} \frac{\sqrt{1-a^2}}{a}$$

$$= \sec^{-1} \frac{1}{\sqrt{1-a^2}} = \operatorname{cosec}^{-1} \frac{1}{a}.$$

$$\cos^{-1} a = \sin^{-1} \sqrt{1-a^2} = \tan^{-1} \frac{\sqrt{1-a^2}}{a} = \cot^{-1} \frac{a}{\sqrt{1-a^2}}$$

$$= \sec^{-1} \frac{1}{a} = \operatorname{cosec}^{-1} \frac{1}{\sqrt{1-a^2}}.$$

$$\tan^{-1} a = \sin^{-1} \frac{a}{\sqrt{1+a^2}} = \cos^{-1} \frac{1}{\sqrt{1+a^2}} = \cot^{-1} \frac{1}{a}.$$

$$= \sec^{-1} \sqrt{1+a^2} = \operatorname{cosec}^{-1} \frac{\sqrt{1+a^2}}{a}.$$

$$\cot^{-1} u = \tan^{-1} \frac{1}{a}; \; \sec^{-1} a = \cos^{-1} \frac{1}{a}; \; \operatorname{cosec}^{-1} a = \sin^{-1} \frac{1}{a}.$$

62   SOLUTION OF TRIGONOMETRIC EQUATIONS

To solve a trigonometric equation, reduce the given equation, by means of the relations expressed in Art. 58 to Art. 60 inclusive, to an equation containing only a single function of a single angle. Solve the resulting equation by algebraic methods (Art. 30) for the remaining function and from this find the values of the angle, by Art. 57 and Art. 59. All these values should then be tested in the original equation, discarding those which do not satisfy it.

63   RELATIONS BETWEEN SIDES AND ANGLES OF PLANE TRIANGLES

Let $a, b,$ and c denote the sides and $\alpha, \beta,$ and $\gamma,$ the corresponding opposite angles with

$2s = a + b + c;$ r = radius of inscribed circle;

A = area; R = radius of circumscribed circle;

h_b = altitude on side b.

$\alpha + \beta + \gamma = 180°.$

$$\frac{a}{\sin \alpha} = \frac{b}{\sin \beta} = \frac{c}{\sin \gamma} \text{ (law of sines).}$$

$$\frac{a+b}{a-b} = \frac{\tan \frac{1}{2}(\alpha + \beta)}{\tan \frac{1}{2}(\alpha - \beta)} \text{ (law of tangents).}*$$

$$a^2 = b^2 + c^2 - 2bc \cos \text{ (law of cosines).}^*$$

$$a = b \cos \gamma + c \cos \beta.^*$$

$$\cos \alpha = \frac{b^2 + c^2 - a^2}{2bc}, * \quad \sin \alpha = \frac{2}{bc}\sqrt{s(s-a)(s-b)(s-c)}.^*$$

$$\sin \frac{\alpha}{2} = \sqrt{\frac{(s-b)(s-c)}{bc}}, * \quad \cos \frac{\alpha}{2} = \sqrt{\frac{s(s-a)}{bc}}; *$$

$$\tan \frac{\alpha}{2} = \sqrt{\frac{(s-b)(s-c)}{s(s-a)}} = \frac{r}{s-a}, * \text{where } r = \sqrt{\frac{(s-a)(s-b)(s-c)}{s}}.$$

$$A = \frac{1}{2}bh_b^* = \frac{1}{2}ab\sin\gamma* \quad = \frac{a^2 \sin\beta \sin\gamma *}{2\sin\alpha}$$

$$= \sqrt{s(s-a)(s-b)(s-c)} = rs.$$

$$R = \frac{a*}{2\sin\alpha} = \frac{abc}{4A}. \quad h_b = c\sin\alpha* = a\sin\gamma* = \frac{2rs}{b}$$

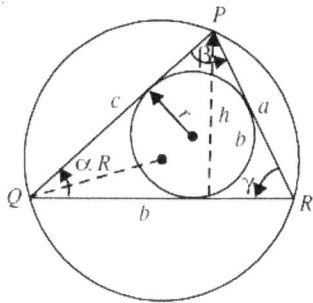

FIGURE 3.2.

64   SOLUTION OF A RIGHT TRIANGLE

Given one side and any acute angle α, or any two sides, to find the remaining parts:

$$a = \sqrt{(c+b)(c-a)} = c \sin \alpha = b \tan \alpha.$$

$$b = \sqrt{(c+a)(c-a)} = c \cos \alpha = \frac{a}{\tan \alpha}.$$

* Two more formulas may be obtained by replacing a by b, b by c, c by a, α by β, β by γ, γ by α.

$$\sin \alpha = \frac{a}{c}, \cos \alpha = \frac{b}{c}, \tan \alpha = \frac{a}{b}, \beta = 90° - \alpha.$$

$$c = \frac{a}{\sin \alpha} = \frac{b}{\cos \alpha} = \sqrt{a^2 + b^2}.$$

$$A = \frac{1}{2}ab = \frac{a^2}{2\tan \alpha} = \frac{b^2 \tan \alpha}{2} = \frac{c^2 \sin 2\alpha}{4}.$$

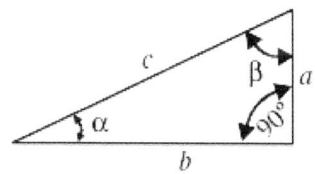

FIGURE 3.3.

65   SOLUTION OF OBLIQUE TRIANGLES

The formulas of the preceding section Art. 63 will suffice to solve any oblique triangle. Use the trigonometric tables for numerical work. We give one method. Solutions should be checked by some other method.

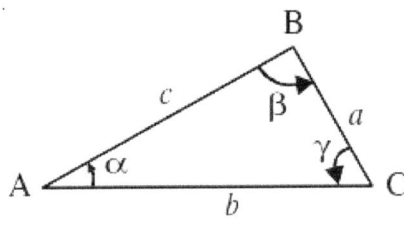

FIGURE 3.4.

(a) Given any two sides b and c and included angle α.

$$\frac{1}{2}(\beta + \gamma) = 90° - \frac{1}{2}\alpha; \tan\frac{1}{2}(\beta - \gamma) = \frac{b - c}{b + c}\tan\frac{1}{2}(\beta + \gamma);$$

$$\beta = \frac{1}{2}(\beta + \gamma) + \frac{1}{2}(\beta - \gamma); \gamma = \frac{1}{2}(\beta + \gamma) - \frac{1}{2}(\beta - \gamma); a = \frac{b\sin\alpha}{\sin\beta}.$$

(b) Given any two angles α and β, and any side c.

$$\gamma = 180° - (\alpha + \beta); a = \frac{c\sin\alpha}{\sin\gamma}; b = \frac{c\sin\beta}{\sin\gamma}.$$

(c) Given any two sides a and c, and an angle opposite one of these, say α.

$$\sin\gamma = \frac{c\sin\alpha}{a}, \beta = 180° - (\alpha + \gamma), b = \frac{a\sin\beta}{\sin\alpha}.$$

This case may have two sets of solutions, for γ may have two values, $\gamma_1 < 90°$ and $\gamma_2 = 180° - \gamma_1 > 90°$. If $\alpha + \gamma_2 > 180°$, use only γ_1.

(d) Given the three sides a, b, and c.

$$s = \frac{1}{2}(a + b + c), r = \sqrt{\frac{(s - a)(s - b)(s - c)}{s}}.$$

$$\tan\frac{1}{2}\alpha = \frac{r}{s - a}, \tan\frac{1}{2}\beta = \frac{r}{s - b}, \tan\frac{1}{2}\gamma = \frac{r}{s - c}.$$

Spherical Trigonometry

66 THE RIGHT SPHERICAL TRIANGLE

Let O be the center of the sphere, and a, b, and c the sides of a right triangle with opposite angles α, β, and $\gamma = 90°$, respectively, the sides being measured by the angle subtended at the center of the sphere.

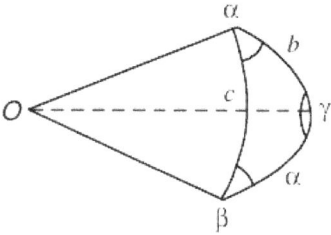

FIGURE 3.5.

$$\sin a = \sin a \cdot \sin c, \qquad \sin b = \sin \beta \cdot \sin c.$$

$$\sin a = \tan b \cdot \cot \beta, \qquad \sin b = \tan a \cdot \cot \alpha.$$

$$\cos \alpha = \cos a \cdot \sin \beta, \qquad \cos \beta = \cos b \cdot \sin \alpha.$$

$$\cos \alpha = \tan b \cdot \cot c, \qquad \cos \beta = \tan a \cdot \cot c.$$

$$\cos c = \cot \alpha \cdot \cot \beta, \qquad \cos c = \cos a \cdot \cos b.$$

67 NAPIER'S RULES OF CIRCULAR PARTS

Let the five quantities a, b, co-α (complement of α), co-c, co-β, be arranged in order as indicated in the figure. If we denote any one of these quantities as a *middle* part, then two of the other parts are *adjacent* to it, and the other two parts are *opposite* to it. The above formulas may be remembered by means of the following rules :

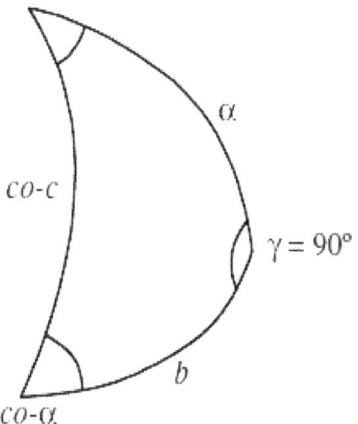

(a) *The sine of a middle part is equal to the product of the tangents of the adjacent parts.*

(b) *The sine of a middle part is equal to the product of the cosines of the opposite parts.*

FIGURE 3.6.

68   THE OBLIQUE SPHERICAL TRIANGLE

Let a, b, and c denote the sides and α, β, and γ the corresponding opposite angles of the spherical triangle, Δ its area, E its spherical excess, R the radius of the sphere upon which the triangle lies, and α', β', γ', a, b, and c the corresponding parts of the polar triangle.

$$0° < a + b + c < 360°, \qquad\qquad 180° < + + < 540°.$$

$$\alpha = 180° - a', \qquad\qquad \beta' = 180° - b', \qquad\qquad \gamma = 180° - c',$$
$$a = 180° - \alpha', \qquad\qquad b = 180° - \beta', \qquad\qquad c = 180° - \gamma'.$$

$$\frac{\sin\alpha}{\sin a} = \frac{\sin\beta}{\sin b} = \frac{\sin\gamma}{\sin c} \quad \text{(law of sines)}.$$

$$\cos a = \cos b \cos c + \sin b \sin c \cos \alpha.$$

$$\cos \alpha = -\cos \beta \cos \gamma + \sin \beta \sin \gamma \cos a \quad \text{(law of cosines)}.$$

$$\tan\frac{\alpha}{2} = \sqrt{\frac{\sin(s-b)\cdot\sin(s-c)}{\sin s \cdot \sin(s-a)}}, \qquad \text{where } s = \frac{1}{2}(a+b+c).$$

$$\tan\frac{a}{2} = \sqrt{\frac{-\cos\sigma\cdot\cos(\sigma-\alpha)}{\cos(\sigma-\beta)\cdot\cos(\sigma-\gamma)}}, \qquad \text{where } \sigma = \frac{1}{2}(\alpha+\beta+\gamma).$$

$$\frac{\sin\frac{1}{2}(\alpha+\beta)}{\sin\frac{1}{2}(\alpha-\beta)} = \frac{\tan\frac{1}{2}c}{\tan\frac{1}{2}(a-b)}, \quad \frac{\cos\frac{1}{2}(\alpha+\beta)}{\cos\frac{1}{2}(\alpha-\beta)} = \frac{\tan\frac{1}{2}c}{\tan\frac{1}{2}(a+b)}.$$

$$\frac{\sin\frac{1}{2}(a+b)}{\sin\frac{1}{2}(a-b)} = \frac{\cot\frac{1}{2}\gamma}{\tan\frac{1}{2}(\alpha-\beta)}, \quad \frac{\cos\frac{1}{2}(a+b)}{\cos\frac{1}{2}(a-b)} = \frac{\cot\frac{1}{2}\gamma}{\tan\frac{1}{2}(\alpha+\beta)}.$$

$$\sin\frac{1}{2}(\alpha+\beta)\cos\frac{1}{2}c = \cos\frac{1}{2}(a-b)\cos\frac{1}{2}\gamma.$$

$$\cos\frac{1}{2}(\alpha+\beta)\cos\frac{1}{2}c = \cos\frac{1}{2}(a+b)\sin\frac{1}{2}\gamma.$$

$$\sin\frac{1}{2}(\alpha-\beta)\sin\frac{1}{2}c = \sin\frac{1}{2}(a-b)\cos\frac{1}{2}\gamma.$$

$$\cos\frac{1}{2}(\alpha-\beta)\sin\frac{1}{2}c = \sin\frac{1}{2}(a+b)\sin\frac{1}{2}\gamma.$$

$$\tan\frac{E}{4} = \sqrt{\tan\frac{1}{2}s\cdot\tan\frac{1}{2}(s-a)\cdot\tan\frac{1}{2}(s-b)\cdot\tan\frac{1}{2}(s-c)}.$$

$$\Delta = \frac{\pi R^2 E}{180}, \qquad \alpha+\beta+\gamma-180° = E.$$

Hyperbolic Functions

69   DEFINITIONS (For definition of *e* see Art. 108)

Hyperbolic sine of $x = \sinh x = \dfrac{1}{2}\left(e^x - e^{-x}\right)$;

Hyperbolic cosine of $x = \cosh x = \dfrac{1}{2}\left(e^x - e^{-x}\right)$;

Hyperbolic tangent of $x = \tanh x = \dfrac{e^x - e^{-x}}{e^x + e^{-x}} = \dfrac{\sinh x}{\cosh x}.$

$$\operatorname{cosech} x = \frac{1}{\sinh x}; \operatorname{sech} x = \frac{1}{\cosh x}; \coth x = \frac{1}{\tanh x}.$$

70   INVERSE OR ANTI-HYPERBOLIC FUNCTIONS

If $x = \sinh y$, is the *inverse hyperbolic sine of x*, write $y = \sinh^{-1} x$ or arc sinh x.

$$\sinh^{-1} x = \log_e\left(x + \sqrt{x^2 + 1}\right), \qquad \cosh^{-1} x = \log_e\left(x + \sqrt{x^2 - 1}\right),$$

$$\tanh^{-1} x = \frac{1}{2}\log_e\left(\frac{1+x}{1-x}\right), \qquad \coth^{-1} x = \frac{1}{2}\log_e\left(\frac{x+1}{x-1}\right),$$

$$\operatorname{sech}^{-1} x = \log_e\left(\frac{1+\sqrt{1-x^2}}{x}\right), \qquad \operatorname{cosech}^{-1} x = \log_e\left(\frac{1+\sqrt{1+x^2}}{x}\right)$$

71 FUNDAMENTAL IDENTITIES

$$\sinh(-x) = -\sinh x, \qquad \cosh^2 x - \sinh^2 x = 1,$$
$$\cosh(-x) = -\cosh x, \qquad \text{sech}^2 x + \tanh^2 x = 1,$$
$$\tanh(-x) = -\tanh x, \qquad \text{cosech}^2 x - \coth^2 x = -1,$$
$$\sinh(x \pm y) = \sinh x \cosh y \pm \cosh x \sinh y.$$
$$\cosh(x \pm y) = \cosh x \cosh y \pm \sinh x \sinh y.$$
$$\tan(x \pm y) = \frac{\tan x \pm \tanh y}{1 \pm \tanh x \tanh y}.$$
$$\sinh 2x = 2 \sinh x \cosh x, \qquad \cosh 2x = \cosh^2 x + \sinh^2 x.$$
$$2 \sinh^2 \frac{x}{2} = \cosh x - 1, \qquad 2 \cosh^2 \frac{x}{2} = \cosh x + 1.$$

72 CONNECTION WITH CIRCULAR FUNCTIONS

$$\sin x = \frac{e^{ix} - e^{-ix}}{2i}, \qquad \cos x = \frac{e^{ix} + e^{-ix}}{2}, \qquad i^2 = -1.$$
$$\sin x = -i \sinh ix, \qquad \cos x = \cosh ix, \qquad \tan x = -i \tanh ix,$$
$$\sin ix = i \sinh ix, \qquad \cos ix = \cosh x \qquad \tan ix = i \tanh ix,$$
$$\sinh ix = i \sin x, \qquad \cosh ix = \cos x, \qquad \tanh ix = i \tan x.$$

$$\sin(x \pm iy) = \sinh x \cos y \pm i \cosh x \sin y.$$
$$\cos(x \pm iy) = \cosh x \cos y \pm i \sinh x \sin y.$$
$$\sinh(x + 2i\pi) = \sinh x, \qquad \cosh(x + 2i\pi) = \cosh x.$$
$$\sinh(x + i\pi) = -\sinh x, \qquad \cosh(x + i\pi) = -\cosh x.$$
$$\sinh\left(x + \frac{1}{2}i\pi\right) = i \cosh x. \qquad \cosh\left(x + \frac{1}{2}i\pi\right) = i \sinh x.$$

$$e^\theta = \cosh \theta + \sinh \theta, \qquad e^{-\theta} = \cosh \theta - \sinh \theta.$$
$$e^{i\theta} = \cosh \theta + i \sinh \theta, \qquad e^{-i\theta} = \cosh \theta - i \sinh \theta.$$
$$x + iy = re^{i\theta}, \text{ where } r = +\sqrt{x^2 + y^2}, \ \theta = \text{arc cos} \frac{x}{r} = \text{arc sin} \frac{y}{r}.$$

$$\log_e (x \pm iy) = \frac{1}{2}\log_e \left(x^2 + y^2\right) \pm i \arctan\frac{y}{x}$$

$$\left(\cos\theta + i\sin\theta\right)^n = \cos n\theta + i\sin n\theta.$$

$$\left(\cos\frac{2k\pi}{n} + i\sin\frac{2k\pi}{n}\right)^n = 1, k = 0, 1, ..., n-1.$$

$$\sinh x + \sinh y = 2\sinh\left(\frac{x+y}{2}\right)\cosh\left(\frac{x-y}{2}\right).$$

$$\sinh x - \sinh y = 2\cosh\left(\frac{x+y}{2}\right)\sinh\left(\frac{x-y}{2}\right).$$

$$\cosh x + \cosh y = 2\cosh\left(\frac{x+y}{2}\right)\cosh\left(\frac{x-y}{2}\right).$$

$$\cosh x - \cosh y = 2\sinh\left(\frac{x+y}{2}\right)\sinh\left(\frac{x-y}{2}\right).$$

CHAPTER

4

ANALYTIC GEOMETRY

73   RECTANGULAR COORDINATES

Let $X'X$ (x-axis) and $Y'Y$
(y-axis) be two perpendicular lines meeting in the
point O called the origin.
The point $P(x, y)$ in the
plane of the x and y axes
is fixed by the distances x
(abscissa) and y (ordinate)
from $Y'Y$ and $X'X$, respectively, to P. x is positive to
the right and negative to
the left of the y-axis, and y
is positive above and negative below the x-axis.

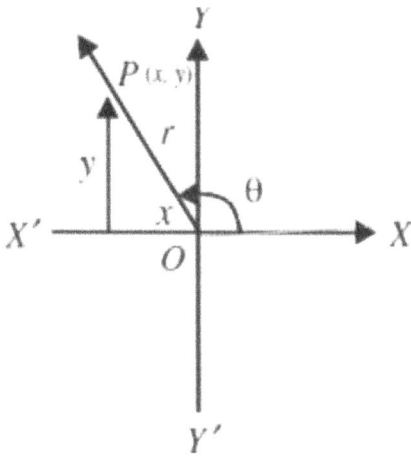

FIGURE 4.1.

74   POLAR COORDINATES

Let OX (initial line) be a fixed line in the plane and O (pole or
origin) a point on this line. The position of any point $P(r, \theta)$ in the
plane is determined by the distance r (radius vector) from O to
the point together with the angle (vectorial angle) measured from
OX to OP. θ is positive if measured counterclockwise, negative if
measured clockwise, r is positive if measured along the terminal
side of θ and negative if measured along the terminal side of produced through the pole.

75 RELATIONS BETWEEN RECTANGULAR AND POLAR COORDINATES

$$\begin{cases} x = r\cos\theta, \\ y = r\sin\theta, \end{cases} \quad \begin{cases} r = \sqrt{x^2+y^2}, \\ \theta = \tan^{-1}\dfrac{y}{x}, \end{cases} \quad \begin{cases} \sin\theta = \dfrac{y}{\sqrt{x^2+y^2}} \\ \cos\theta = \dfrac{x}{\sqrt{x^2+y^2}} \end{cases}$$

76 POINTS AND SLOPES

Let $P_1(x_1, y_1)$ and $P_2(x_2, y_2)$ be any two points, and α_1 be the angle measured counterclockwise from OX to P_1P_2:

Distance between P_1 and $P_2 = P_1P_2$

$$d = \sqrt{(x_2 - x_1)^2 + (y_2 - y_1)^2}.$$

Slope of $P_1P_2 = \tan\alpha_1 = m = \dfrac{y_2 - y_1}{x_2 - x_1}.$

Point dividing P_1P_2 in the ratio $m_1 : m_2$ is $\left(\dfrac{m_1x_2 + m_2x_1}{m_1 + m_2}, \dfrac{m_1y_2 + m_2y_1}{m_1 + m_2} \right)$

Mid-point of $P_1 P_2$ is $\left(\dfrac{x_1 + x_2}{2}, \dfrac{y_1 + y_2}{2} \right).$

The angle β between lines of slopes m_1 and m_2, respectively, is given by

$$\tan\beta = \dfrac{m_2 - m_1}{1 + m_1m_2}.$$

Two lines of slopes m_1 and m_2 are perpendicular if $m_2 = -\dfrac{1}{m_1}$, and parallel if $m_1 = m_2$.

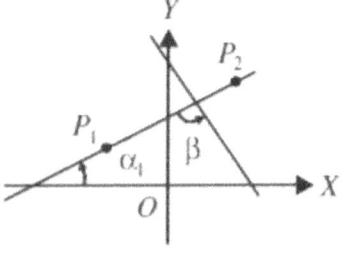

FIGURE 4.2.

77   AREA OF TRIANGLE

If the vertices are the points (x_1, y_1), (x_2, y_2), and (x_3, y_3), then the area is equal to the numerical value of

$$\frac{1}{2}\begin{vmatrix} x_1 & y_1 & 1 \\ x_2 & y_2 & 1 \\ x_3 & y_3 & 1 \end{vmatrix} = \frac{1}{2}\left(x_1 y_2 - x_1 y_3 + x_2 y_3 - x_2 y_1 + x_3 y_1 - x_3 y_2\right).$$

78   LOCUS AND EQUATION

The set of all points which satisfy a given condition is called the *locus* of that condition. An *equation* is called the equation of the locus if it is satisfied by the coordinates of every point on the locus and by no other points. There are three common representations of the locus by means of equations:

(a) *Rectangular equation* which involves the rectangular coordinates (x, y).

(b) *Polar equation* which involves the polar coordinates (r, θ).

(c) *Parametric equations* which express x and y or r and θ in terms of a third independent variable called a parameter.

79   TRANSFORMATION OF COORDINATES

To transform an equation or a curve from one system of coordinates in x, y to another such system in x', y', substitute for each variable its value in terms of variables of the new system.

(a) *Rectangular System: Old axes parallel to new axes.*

The coordinates of new origin in terms of the old system are (h, k).

$$\begin{cases} x &=& x' + h, \\ y &=& y' + k \end{cases}$$

(b) *Rectangular System: Old origin coincident with a new origin and the x´-axis making an angle θ with the x-axis.*

$$\begin{cases} x &=& x' \cdot \cos\theta - y' \cdot \sin\theta. \\ y &=& x' \cdot \sin\theta + y' \cdot \cos\theta. \end{cases}$$

(c) *Rectangular System: Old axes are not parallel with new ones. New origin at* (h, k) *in the old system.*

$$\begin{cases} x &=& x' \cdot \cos\theta - y' \cdot \sin\theta + h. \\ y &=& x' \cdot \sin\theta + y' \cdot \cos\theta + k. \end{cases}$$

80 STRAIGHT LINE

The equations of the straight line may assume the following forms:

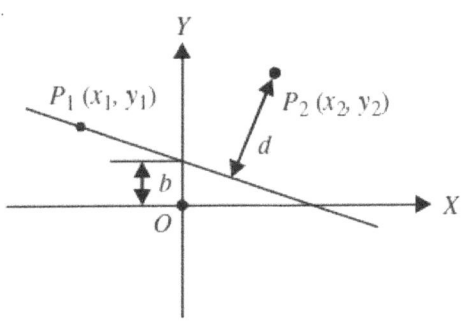

FIGURE 4.3.

(a) $y = mx + b.$ ($m =$ slope, $b =$ intercept on y-axis).

(b) $y - y_1 = m(x - x_1).$ [$m =$ slope, line passes through point (x_1, y_1)].

(c) $\dfrac{y - y_1}{x - x_1} = \dfrac{y_2 - y_1}{x_2 - x_1}.$ [line passes thru points (x_1, y_1) and (x_2, y_2)].

(d) $\dfrac{x}{a} + \dfrac{y}{b} = 1.$ (a and b are the intercepts on the x and y-axis, respectively).

(e) $x \cos \alpha + y \sin \alpha = p.$ (*normal form*, p is distance from origin to the line, α is angle which normal to the line makes with x-axis).

(f) $Ax + By + C = 0.$ (*general form*, slope $= -A \div B$).

To reduce $Ax + By + C = 0$ to normal form (e), divide by $\pm \sqrt{A^2 + B^2}$, where the sign of the radical is taken opposite to that of C when $C \neq 0$.

The distance from the line $Ax + By + C = 0$ to the point $P_2(x_2, y_2)$ is:

$$d = \frac{Ax_2 + By_2 + C}{\pm\sqrt{A^2 + B^2}}.$$

81   CIRCLE

General equation of a circle with radius R and center at (h, k) is:

$(x - h)^2 + (y - k)^2 = R^2$.

82   CONIC

The locus of a point P which moves so that its distance from a fixed Point F (focus) bears a constant ratio e (eccentricity) its distance from a fixed straight line (directrix) is a conic.

If d is the distance from focus to directrix, and F is at the origin.

$$x^2 + y^2 = e^2(d + x)^2.$$

$$r = \frac{de}{1 - e\cos\theta}.$$

If $e = 1$, the conic is a *parabola*; if

$e > 1$, a *hyperbola*; and if $e < 1$, an *ellipse*.

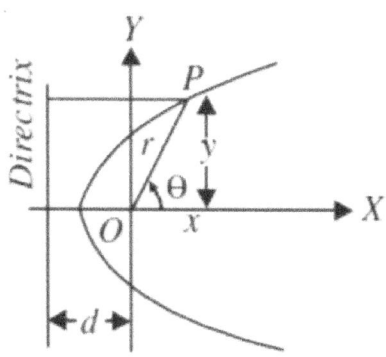

FIGURE 4.4.

83   PARABOLA

$e = 1.$

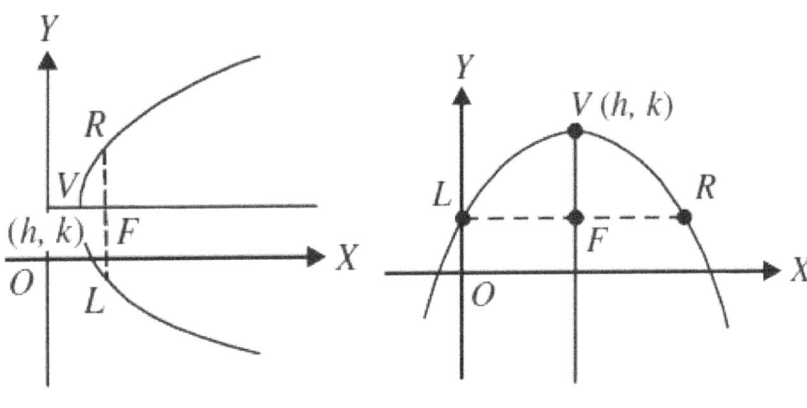

FIGURE 4.5.   FIGURE 4.6.

(a) $(y - k)^2 = 4a(x - h)$. Vertex at (h, k), axis $\parallel OX$ (Figure 4.5).

Figure 4.5 is drawn for the case when a is positive.

(b) $(x - h)^2 = 4a(y - k)$. Vertex at (h, k) axis $\parallel OY$ (Figure 4.6).

Figure 4.6 is drawn for the case when a is negative.

Distance from vertex to focus $= VF = a$.

Distance from vertex to directrix $= a$.

Latus rectum $= LR = 4a$.

84   ELLIPSE

$e < 1.$

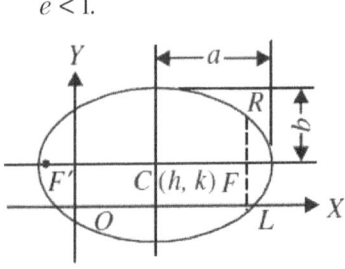

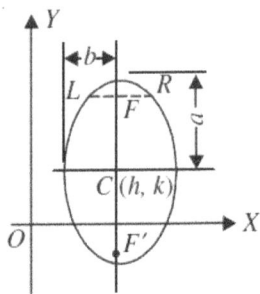

FIGURE 4.7.   FIGURE 4.8.

(a) $\dfrac{(x-h)^2}{a^2}+\dfrac{(y-k)^2}{b^2}=1$. Center at (h, k), major axis $\parallel OX$ (Figure 4.7).

(b) $\dfrac{(y-k)^2}{a^2}+\dfrac{(x-h)^2}{b^2}=1$. Center at (h, k), major axis $\parallel OY$ (Figure 4.8).

Major axis = $2a$ Major axis = $2b$.

Distance from center to either focus $=\sqrt{a^2-b^2}$.

Distance from center to either directrix $= a/e$.

Eccentricity $= e = \sqrt{a^2-b^2}\,/\,a$.

Latus rectum $= 2b^2/a$.

Sum of distances from any point P on the ellipse to foci, $PF' + PF = 2a$.

85 HYPERBOLA

$e > 1$.

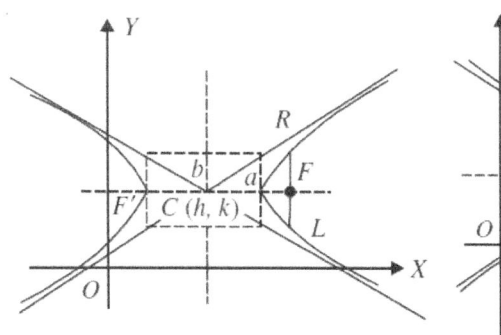

FIGURE 4.9.

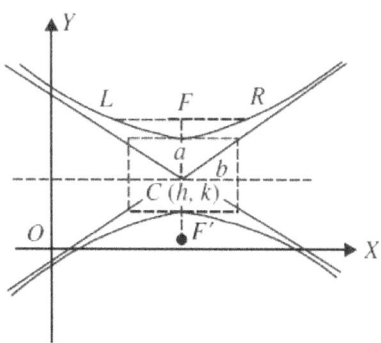

FIGURE 4.10.

(a) $\dfrac{(x-h)^2}{a^2}-\dfrac{(y-k)^2}{b^2}=1$.

Center at (h, k), transverse axis $\parallel OX$.

Slopes of asymptotes $= \pm\, b/a$ (Figure 4.9).

(b) $\dfrac{(y-k)^2}{a^2} - \dfrac{(x-h)^2}{b^2} = 1.$

Center at (h, k), transverse axis $\parallel OY$.

Slopes of asymptotes $= \pm\, a/b$ (Figure 4.10).

Transverse axis $= 2a$.

Conjugate axis $= 2b$.

Distance from center to either focus $= \sqrt{a^2 + b^2}$.

Distance from center to either directrix $= a/e$.

Difference of distances of any point on hyperbola from the foci $= 2a$.

Eccentricity $= e = \dfrac{\sqrt{a^2 + b^2}}{a}.$

Latus Rectum $= 2b^2/a$.

86   SINE CURVE

$y = a \sin (bx = c).$

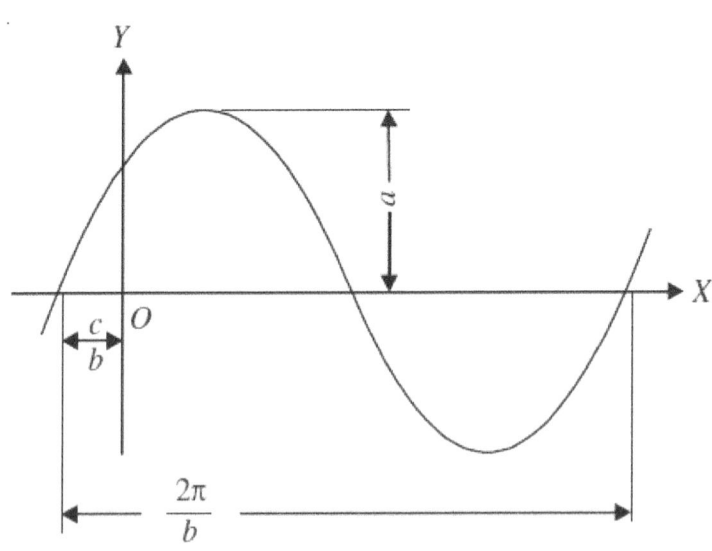

FIGURE 4.11.

$y = a \cos (bx + c\,') = a \sin (bx + c),$ where $c = c\,' + \dfrac{\pi}{2}.$

$y = p \sin bx + q \cos bx = a \sin (bx + c),$

where $c = \tan^{-1}\left(\dfrac{q}{p}\right), a = \sqrt{p^2 + q^2}.$

a = amplitude = maximum height of wave.

$\dfrac{2\pi}{b}$ = wave length = distance from any point on wave to the corresponding point on the next wave.

$x = -\dfrac{c}{b}$ = phase, indicates a point on x-axis from which the positive half of the wave starts.

87   TRIGONOMETRIC CURVES

(1) $y = a \tan bx$, or

$x = \dfrac{1}{b}\tan^{-1}\left(\dfrac{y}{a}\right).$

(1) $y = a \sec bx$, or,

$x = \dfrac{1}{b}\sec^{-1}\left(\dfrac{y}{a}\right).$

(2) $y = a \cot bx$, or,

$x = \dfrac{1}{b}\cot^{-1}\left(\dfrac{y}{a}\right).$

(2) $y = a \operatorname{cosec} bx$, or,

$x = \dfrac{1}{b}\operatorname{cosec}^{-1}\left(\dfrac{y}{a}\right).$

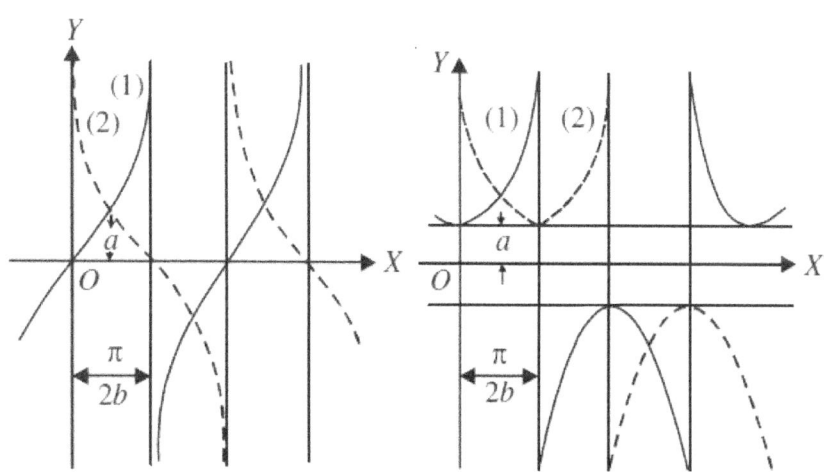

FIGURE 4.12.   FIGURE 4.13.

88 LOGARITHMIC AND EXPONENTIAL CURVES

$y = \log_a x$ or $x = a^y$.

Logarithmic Curve

$y = a_x$ or $x = \log_a y$.

Exponential Curve

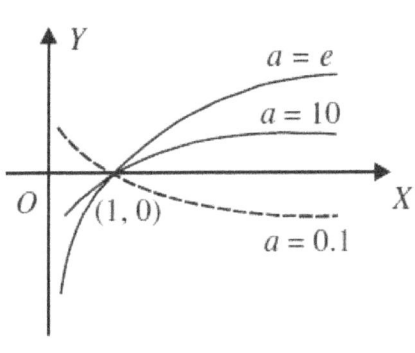

FIGURE 4.14

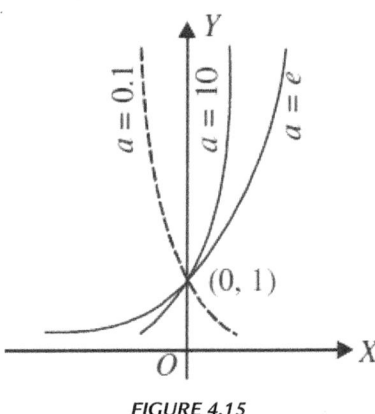

FIGURE 4.15

89 PROBABILITY CURVE (FIGURE 4.16)

$y = e^{-x^2}$

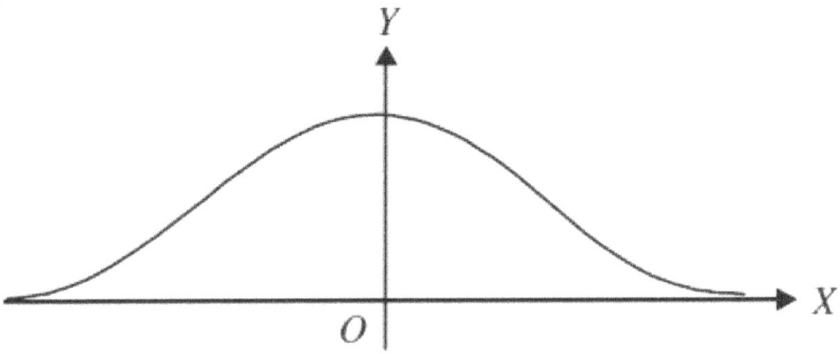

FIGURE 4.16.

90 OSCILLATORY WAVE OF DECREASING AMPLITUDE (FIGURE 4.17)

$y = e^{-ax} \sin bx$.

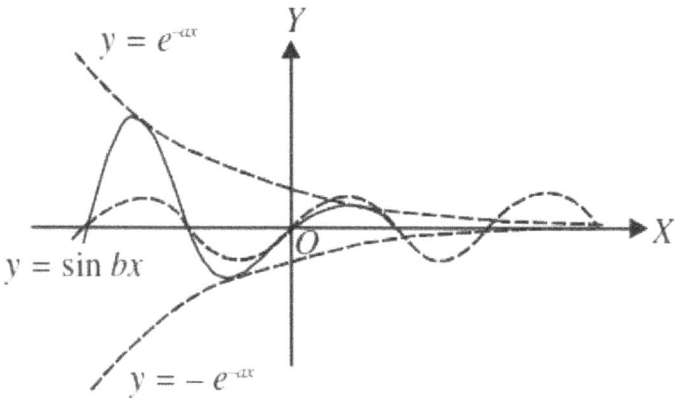

FIGURE 4.17.

91   CATENARY (FIGURE 4.18)

$$y = \frac{a}{2}\left(e^{\frac{x}{a}} + e^{-\frac{x}{a}} \right).$$

A curve made by a cord of uniform weight suspended freely between two points.

Length of

$$\text{arc} = s = l\left[1 + \frac{2}{3}\left(\frac{2d^2}{l} \right)^2 \right],$$

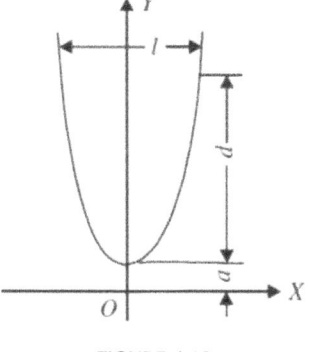

FIGURE 4.18.

approximately, if d is small in comparison with l.

92   CYCLOID

$$\begin{cases} x & = & a(\phi - \sin\phi). \\ y & = & a(1 - \cos\phi). \end{cases}$$

A curve is described as a point on the circumference of a circle which rolls along a fixed straight line.

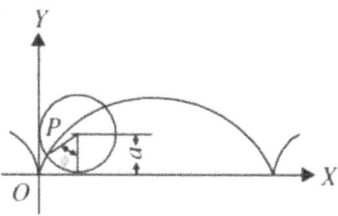

FIGURE 4.19.

Area one arch = $3\pi a^2$.

Length of the arc of one arch = $8a$.

93   PROLATE AND CURTATE CYCLOID

$$\begin{cases} x & = & a\phi - b\sin\phi, \\ y & = & a - b\cos\phi. \end{cases}$$

A curve is described by a point on a circle at a distance b from the center of the circle of radius a which rolls along a fixed straight line.

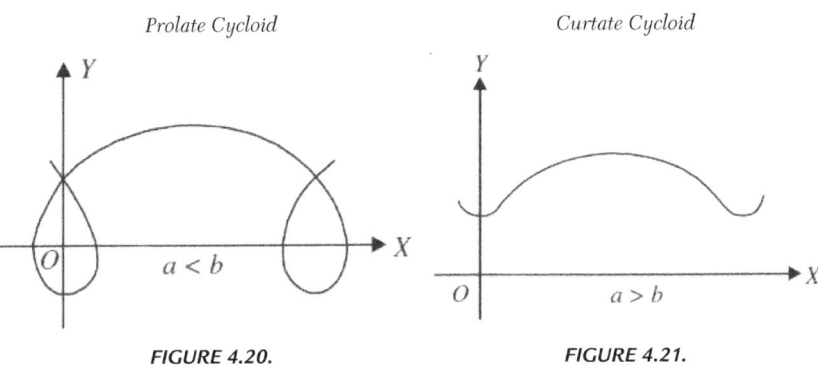

Prolate Cycloid   *Curtate Cycloid*

FIGURE 4.20.   **FIGURE 4.21.**

94   EPICYCLOID

$$\begin{cases} x & = & (a+b)\cos\phi - a\cos\left(\dfrac{a+b}{a}\phi\right), \\ y & = & (a+b)\sin\phi - a\sin\left(\dfrac{a+b}{a}\phi\right). \end{cases}$$

A curve is described as a point on the circumference of a circle that rolls along the outside of a fixed circle.

95   CARDIOID

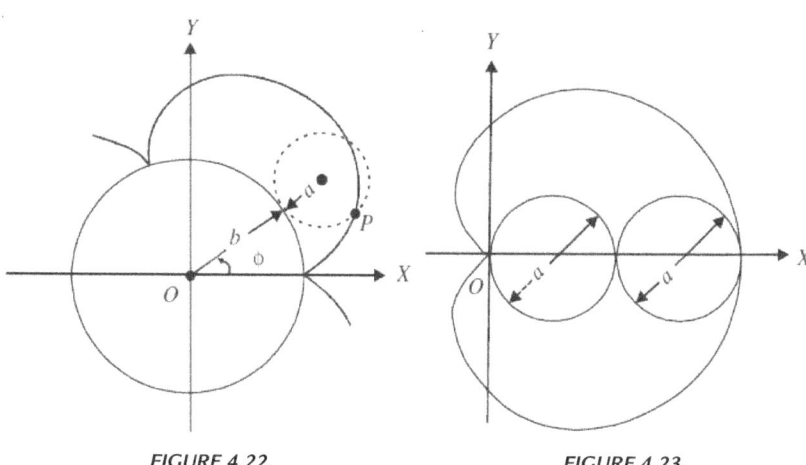

FIGURE 4.22.   FIGURE 4.23.

96   CARDIOID

$$r = a(1 + \cos\theta).$$

An epicycloid in which both circles have the same radii is called a cardioid.

97   HYPOCYCLOID

$$\begin{cases} x &= (a-b)\cos\phi - b\cos\left(\dfrac{a-b}{b}\phi\right), \\ y &= (a-b)\sin\phi - b\sin\left(\dfrac{a-b}{b}\phi\right) \end{cases}$$

A curve is described as a point on the circumference of a circle that rolls along the inside of a fixed circle.

98   HYPOCYCLOID OF FOUR CUSPS

$$x^{\frac{2}{3}} + y^{\frac{2}{3}} = a^{\frac{2}{3}}.$$

$$x = a\cos^3\phi, \qquad\qquad y = a\sin^3\phi.$$

The radius of a fixed circle is four times the radius of a rolling circle.

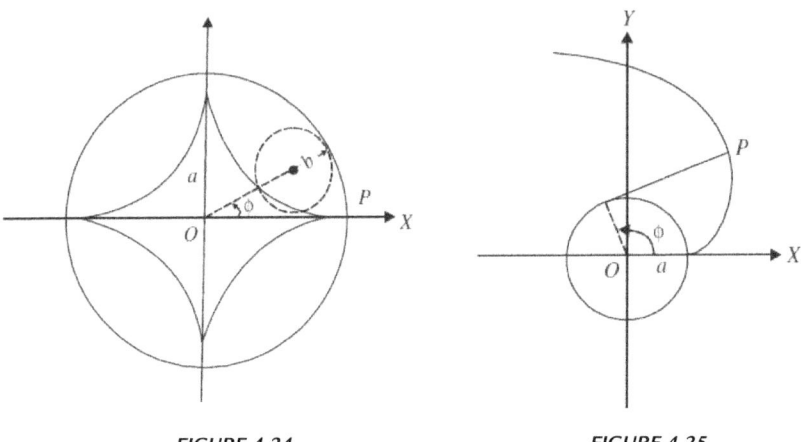

FIGURE 4.24.   FIGURE 4.25.

99   INVOLUTE OF THE CIRCLE

$$\begin{cases} x &=& a\cos\phi + a\,\phi\sin\phi \\ y &=& a\sin\phi + a\,\phi\cos\phi. \end{cases}$$

A curve generated by the end of a string is kept taut while being unwound from a circle.

100   LEMNISCATE

$$r^2 = 2\,a^2 \cos 2\theta.$$

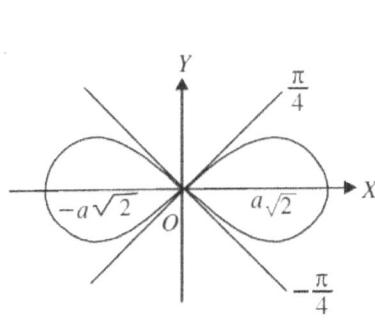

FIGURE 4.26.

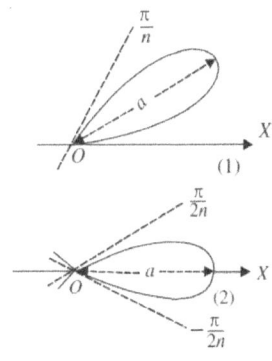

FIGURE 4.27.

101 N-LEAVED ROSE

(1) $r = a \sin n\theta$ (2) $r = a \cos n\theta$.

If n is an odd integer there are n leaves, and if n is even, $2n$ leaves, of which the figure shows one.

102 SPIRALS

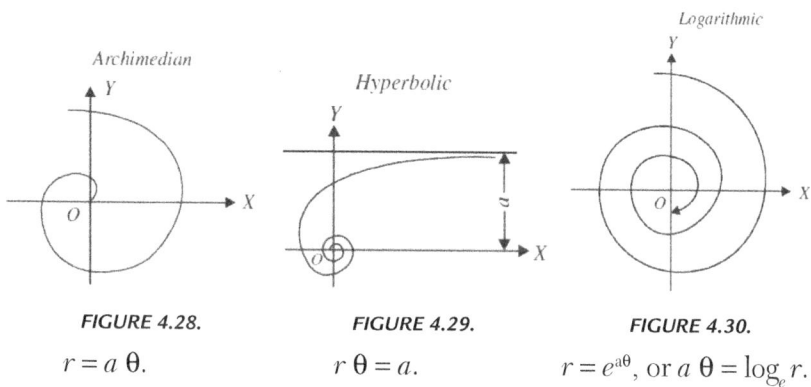

FIGURE 4.28.	FIGURE 4.29.	FIGURE 4.30.
$r = a\,\theta$.	$r\,\theta = a$.	$r = e^{a\theta}$, or $a\,\theta = \log_e r$.

Solid Analytic Geometry

103 COORDINATES (FIGURE 4.31)

(a) Rectangular system: The position of a point $P(x, y, z)$ in space is fixed by its three distances x, y, and z from three coordinate planes XOY, XOZ, and ZOY, which are mutually perpendicular and meet in a point O (origin).

(b) Cylindrical system: The position of any point $P(r, \theta, z)$ is fixed by (r, θ), the polar coordinates of the projection of P in the XOY plane, and by z, its distance from the XOY plane.

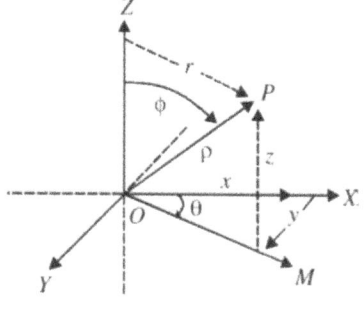

FIGURE 4.31.

(c) Spherical (or polar or geographical) system: The position of any point $P(\rho, \theta, \phi)$ is fixed by the distance $\rho = \overline{OP}$, the angle $\theta = \angle XOM$ and the angle $\phi = \angle ZOP$. θ is called the *longitude* and the *co-latitude*.

The following relations exist between the three coordinates systems:

$$\begin{cases} x = \rho \sin\phi \cos\theta, \\ y = \rho \sin\phi \sin\theta, \\ z = \rho \cos\phi. \end{cases} \quad \begin{cases} r = \rho \sin\phi, \\ z = \rho \cos\phi, \end{cases} \quad \begin{cases} \rho = \sqrt{r^2 + z^2}. \\ \cos\phi = \dfrac{z}{\sqrt{r^2 + z^2}}. \end{cases}$$

104   POINTS, LINES, AND PLANES

Distance between two points $P_1(x_1, y_1, z_1)$ and $P_2(x_2, y_2, z_2)$, is

$$d = \sqrt{(x_2 - x_1)^2 + (y_2 - y_1)^2 + (z_2 - z_1)^2}.$$

Direction cosines of a line are the cosines of the angles α, β, and γ which the line or any parallel line makes with the coordinate axes.

The direction cosines of the line segment $P_1 (x_1, y_1, z_1)$ to $P_2(x_2, y_2, z_2)$ are:

$$\cos\alpha = \frac{x_2 - x_1}{d}, \qquad \cos\beta = \frac{y_2 - y_1}{d}, \qquad \cos\gamma = \frac{z_2 - z_1}{d}.$$

$$\cos^2\alpha + \cos^2\beta + \cos^2\gamma = 1.$$

Angle θ between two lines, whose direction angles are α_1, β_1, γ_1, and α_2, β_2, γ_2, is given by

$$\cos\theta = \cos\alpha_1 \cos\alpha_2 + \cos\beta_1 \cos\beta_2 + \cos\gamma_1 \cos\gamma_2.$$

The equation of a plane is:

$$Ax + By + Cz + D = 0,$$

where A, B, and C are proportional to the direction cosines of a normal (a line perpendicular to the plane) to the plane.

Angle between two planes is the angle between their normals.

Equation of a straight line through the point $P_1(x_1, y_1, z_1)$ is:

$$\frac{x - x_1}{a} = \frac{y - y_1}{b} = \frac{z - z_1}{c},$$

where a, b, and c are proportional to the direction cosines of the line. a, b, *and* c are called the *direction numbers* of the line.

105 FIGURES IN THREE DIMENSIONS

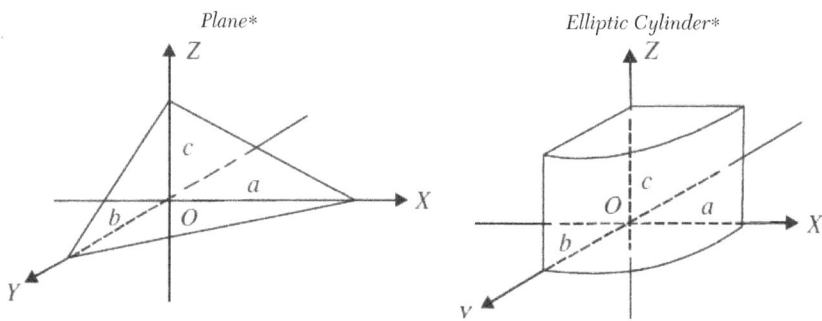

Plane*

FIGURE 4.32.

Elliptic Cylinder*

FIGURE 4.33.

$$\frac{x}{a} + \frac{y}{b} + \frac{z}{c} = 1.$$

$$\frac{x^2}{a^2} + \frac{y^2}{b^2} = 1.$$

* Only a portion of the figure is shown.

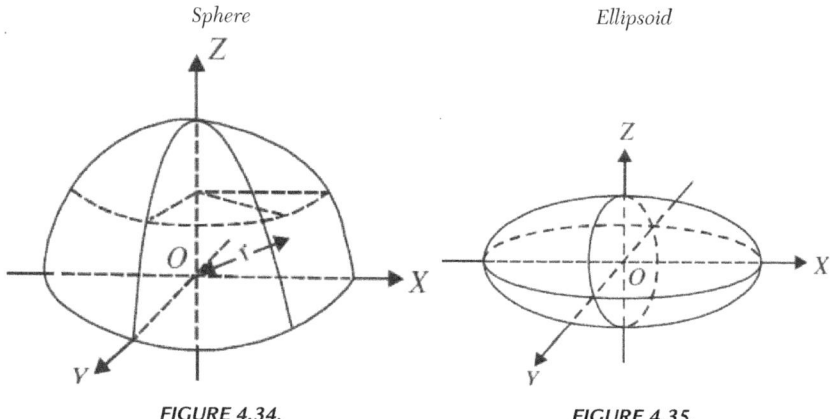

Sphere

FIGURE 4.34.

Ellipsoid

FIGURE 4.35.

$$x^2 + y^2 + z^2 = r^2.$$

$$\frac{x^2}{a^2} + \frac{y^2}{b^2} + \frac{z^2}{c^2} = 1.$$

Elliptic Paraboloid *Portion of Cone* *Hyperboloid of One Sheet*

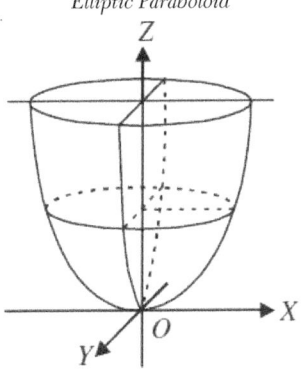
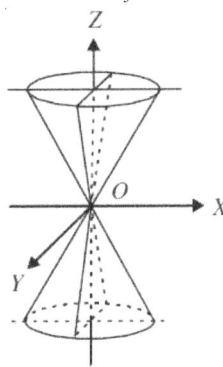
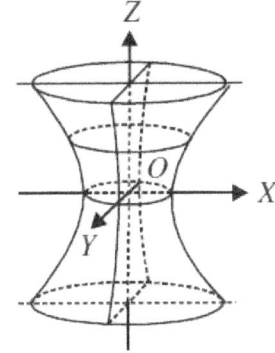

FIGURE 4.36. **FIGURE 4.37.** **FIGURE 4.38.**

$$\frac{x^2}{a^2} + \frac{y^2}{b^2} = cz. \qquad \frac{x^2}{a^2} + \frac{y^2}{b^2} - \frac{z^2}{c^2} = 0. \qquad \frac{x^2}{a^2} + \frac{y^2}{b^2} - \frac{z^2}{c^2} = 1.$$

Hyperboloid of Two Sheets *Hyperbolic Paraboloid*

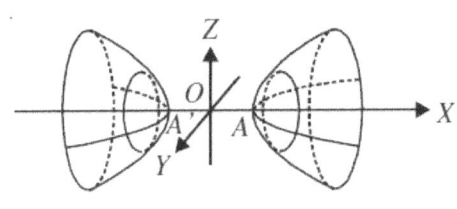
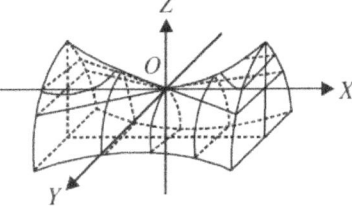

FIGURE 4.39. **FIGURE 4.40.**

$$\frac{x^2}{a^2} - \frac{y^2}{b^2} - \frac{z^2}{c^2} = 1. \qquad \frac{x^2}{a^2} - \frac{y^2}{b^2} = cz.$$

5

DIFFERENTIAL CALCULUS

106   DEFINITION OF FUNCTION

A variable y is said to be a *function* of the variable x, if, for every x (taken at will on its range), y is determined. The symbols $f(x)$, $F(x)$, $g(x)$, $\phi(x)$, etc., are used to represent various functions of x. The symbol $f(a)$ represents the value of $f(x)$ when $x = a$.

107   DEFINITION OF DERIVATIVE AND NOTATION

Let $y = f(x)$ be a single-valued (continuous) function of x. Let Δx be any increment (increase or decrease) given to x, and let Δy be the corresponding increment in y. The *derivative* of y with respect to x is the limit, (if it exists), of the ratio of Δy to Δx as Δx approaches zero in any manner whatsoever; that is,

$$\frac{dy}{dx} = \lim_{\Delta x \to 0} \frac{\Delta y}{\Delta x} = \lim_{\Delta x \to 0} \frac{f(x + \Delta x) - f(x)}{\Delta x} = f'(x) = D_x y = y'.$$

Higher derivatives are defined as follows:

$$\frac{d^2 y}{dx^2} = \frac{d}{dx}\left(\frac{dy}{dx}\right) = \frac{d}{dx} f'(x) = f''(x). \qquad \text{(second derivative)}$$

$$\frac{d^3 y}{dx^3} = \frac{d}{dx}\left(\frac{d^2 y}{dx^2}\right) = \frac{d}{dx} f''(x) = f'''(x). \qquad \text{(third derivative)}$$

$$\frac{d^n y}{dx^n} = \frac{d}{dx}\left(\frac{d^{n-1}y}{dx^{n-1}}\right) = \frac{d}{dx} f^{(n-1)}(x) = f^{(n)}(x). \qquad \text{(nth derivative)}$$

The symbol $f^{(n)}(a)$ represents the value of $f^{(n)}(x)$ when $x = a$.

108   CERTAIN RELATIONS AMONG DERIVATIVES

If $x = f(y)$,   then $\dfrac{dy}{dx} = 1 \div \dfrac{dx}{dy}$.

If $y = f(u)$, and $u = F(x)$, then $\dfrac{dy}{dx} = \dfrac{dy}{du} \cdot \dfrac{du}{dx}$

If $x = f(\alpha),\, y = \phi(\alpha)$,   then

$$\frac{dy}{dx} = \frac{\phi'(\alpha)}{f'(\alpha)},\ \frac{d^2 y}{dx^2} = \frac{f'(\alpha) \cdot \phi''(\alpha) - \phi'(\alpha) \cdot f''(\alpha)}{\left[f'(\alpha)\right]^3}.$$

109   TABLE OF DERIVATIVES

In this table, u and v represent functions of x: a, n, and e represent constants ($e = 2.7183\ ...$), and all angles are measured in radians:

$$\frac{d}{dx}(x) = 1 \qquad\qquad \frac{d}{dx}(a) = 0.$$

$$\frac{d}{dx}(u \pm v \pm ...) = \frac{du}{dx} \pm \frac{dv}{dx} \pm ....$$

$$\frac{d}{dx}(au) = a\frac{du}{dx}. \qquad\qquad \frac{d}{dx}(uv) = u\frac{dv}{dx} + v\frac{du}{dx}.$$

$$\frac{d}{dx}\left(\frac{u}{v}\right) = \frac{v\dfrac{du}{dx} - u\dfrac{dv}{dx}}{v^2}. \qquad\qquad \frac{d}{dx}\sin u = \cos u\frac{du}{dx}.$$

$$\frac{d}{dx}\left(u^n\right) = nu^{n-1}\frac{du}{dx}. \qquad\qquad \frac{d}{dx}\cos u = -\sin u\frac{du}{dx}.$$

$$\frac{d}{x}\log_a u = \frac{\log_a e}{u}\frac{du}{dx}.$$

$$\frac{d}{dx}\tan u = \sec^2 u\frac{du}{dx}.$$

$$\frac{d}{dx}\log_e u = \frac{1}{u}\frac{du}{dx}.$$

$$\frac{d}{dx}\cot u = -\csc^2 u\frac{du}{dx}.$$

$$\frac{d}{dx}a^u = a^u \cdot \log_e a \cdot \frac{du}{dx}.$$

$$\frac{d}{dx}\sec u = \sec u\tan u\frac{du}{dx}.$$

$$\frac{d}{dx}e^u = e^u\frac{du}{dx}.$$

$$\frac{d}{dx}\csc u = -\csc u\cot u\frac{du}{dx}.$$

$$\frac{d}{dx}u^v = vu^{v-1}\frac{du}{dx} + u^v\log_e u\frac{dv}{dx}$$

$$\frac{d}{dx}\text{vers } u = \sin u\frac{du}{dx}.$$

$$\lim_{x\to 0}\frac{\sin x}{x} = 1,\ \lim_{x\to 0}(1+x)^{1/x} = e = 2.71828\ldots = 1 + 1 + \frac{1}{2!} + \frac{1}{3!} + \cdots.$$

$$\frac{d}{dx}\sin^{-1} u = \frac{1}{\sqrt{1-u^2}}\frac{du}{dx},$$

$$\frac{-\pi}{2} \leq \sin^{-1} u \leq \frac{\pi}{2}.$$

$$\frac{d}{dx}\cos^{-1} u = \frac{1}{\sqrt{1-u^2}}\frac{du}{dx},$$

$$0 \leq \cos^{-1} u \leq \pi.$$

$$\frac{d}{dx}\tan^{-1} u = \frac{1}{1+u^2}\frac{du}{dx}.$$

$$\frac{d}{dx}\cot^{-1} u = -\frac{1}{1+u^2}\frac{du}{dx}.$$

$$\frac{d}{dx}\sec^{-1} u = \frac{1}{u\sqrt{u^2-1}}\frac{du}{dx},\ -\pi \leq \sec^{-1} u < -\frac{\pi}{2},\ 0 \leq \sec^{-1} u < \frac{\pi}{2}.$$

$$\frac{d}{dx}\csc^{-1} = -\frac{1}{u\sqrt{u^2-1}}\frac{du}{dx},\ -\pi < \csc^{-1} u \leq -\frac{\pi}{2},\ 0 < \csc^{-1} u \leq \frac{\pi}{2}.$$

$$\frac{d}{dx}\text{vers}^{-1}u = \frac{1}{\sqrt{2u-u^2}}\frac{du}{dx},v$$

$$0 \leq \text{vers}^{-1}u \leq \pi.$$

$$\frac{d}{dx}\sinh u = \cosh u\frac{du}{dx}.$$

$$\frac{d}{dx}\cosh u = \sinh u\frac{du}{dx}.$$

$$\frac{d}{dx}\tanh u = \operatorname{sech}^2 u \frac{du}{dx}.$$

$$\frac{d}{dx}\coth u = -\operatorname{cosech}^2 u \frac{du}{dx}.$$

$$\frac{d}{dx}\operatorname{sech} u = -\operatorname{sech} u \tanh u \frac{du}{dx}.$$

$$\frac{d}{dx}\operatorname{cosech} u = -\operatorname{cosech} u \coth u \frac{du}{dx}.$$

$$\frac{d}{dx}\sinh^{-1} u = \frac{1}{\sqrt{u^2+1}}\frac{du}{dx}.$$

$$\frac{d}{dx}\cosh^{-1} u = \frac{1}{\sqrt{u^2-1}}\frac{du}{dx}, u > 1.$$

$$\frac{d}{dx}\tanh^{-1} u = \frac{1}{1-u^2}\frac{du}{dx}.$$

$$\frac{d}{dx}\coth^{-1} u = -\frac{1}{u^2-1}\frac{du}{dx}.$$

$$\frac{d}{dx}\operatorname{sech}^{-1} u = -\frac{1}{u\sqrt{1-u^2}}\frac{du}{dx}, u > 0.$$

$$\frac{d}{dx}\operatorname{cosech}^{-1} u = -\frac{1}{u\sqrt{u^2+1}}\frac{du}{dx}.$$

110 SLOPE OF A CURVE: EQUATION OF TANGENT AND NORMAL (RECTANGULAR COORDINATES)

The slope of the curve $y = f(x)$ at the point $P(x, y)$ is defined as the slope of the tangent line to the curve at P:

$$\text{Slope} = m = \tan \alpha = \frac{dy}{dx} = f'(x).$$

Slope at $x = x_1$ is $m_1 = f'(x_1)$.

Equation of *tangent line* to curve at $P_1 (x_1, y_1)$ is

$$y - y_1 = m_1 (x - x_1).$$

Equation of *normal line* to curve at $P_1 (x_1, y_1)$ is

$$y - y_1 = -\frac{1}{m_1}(x - x_1).$$

Angle (θ) *of intersection of two curves* whose slopes are m_1 and m_2 at the common point is given by

$$\tan \theta = \frac{m_2 - m_1}{1 + m_1 m_2}.$$

The sign of $\tan \theta$ determines whether the acute or obtuse angle is meant.

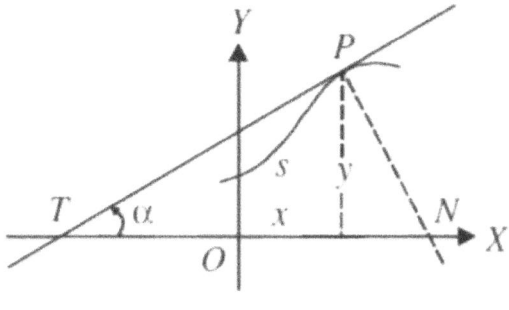

FIGURE 5.1

111 DIFFERENTIAL

If $y = f(x)$, Δx is an increment of x, and $f'(x)$ is the derivative of $f(x)$ with respect to x, then the differential of x equals the increment of x, or $dx = \Delta x$; and the differential of y, dy is the product of $f'(x)$ and the increment of x;

$$dy = f'(x)dx = \frac{df(x)}{dx}dx = \frac{dy}{dx}dx, \quad \text{and} \quad f'(x) = \frac{dy}{dx}.$$

If $x = f(t)$, $y = \phi(t)$, then $dx = f'(t)\, dt$ and $dy = \phi'(t)\, dt$.

Every derivative formula has a corresponding differential formula.

$d(\sin u) = \cos u \cdot du;$ $\qquad d(u \cdot v) = u\, dv + v\, du.$

112 MAXIMUM AND MINIMUM VALUES OF A FUNCTION

A maximum (minimum) value of a function $f(x)$ in the interval (a, b) is a value of the function which is greater (less) than the values of the function in the immediate vicinity.

The values of x which give a maximum or minimum value to $y = f(x)$ are found by solving the equations $f'(x) = 0$ or ∞. If a is a root of $f'(x) = 0$ and if $f''(a) < 0$, $f(a)$ is a maximum; if $f''(a) > 0$, $f(a)$ is a minimum. If $f''(a) = 0$, $f'''(a)$ 0, $f(a)$ is neither a maximum nor a minimum, but if $f''(a) = f'''(a) = 0$, $f(a)$ is maximum or minimum according as $f^{IV}(a) \lessgtr 0$.

In general, if the first derivative which does not vanish for $x = a$ is of odd order, $f(a)$ is neither a maximum nor a minimum; but if it is of even order, the $2nth$, say, then $f(a)$ is a maximum or minimum according as $f^{(2n)}(a) \lessgtr 0$.

To find the largest or smallest values of a function in an interval (a, b), find $f(a)$, $f(b)$, and compare with the maximum and minimum values as found in the interval.

113   POINTS OF INFLECTION OF A CURVE

The curve is said to have a *point of inflection* at $x = a$ if $f''(a) = 0$ and $f''(x) < 0$ on one side of $x = a$ and $f''(x) > 0$ on the other side of $x = a$. Wherever $f''(x) < 0$, the curve is *concave downward*, and wherever $f''(x) > 0$, the curve is *concave upward*.

114   DERIVATIVE OF ARC LENGTH: RADIUS OF CURVATURE

Let s be the length of arc measured along the curve $y = f(x)$, [or in polar coordinates $r = \phi(\theta)$], from some fixed point to any point $P(x, y)$, and α be the angle of inclination of the tangent line at P with OX. Then

$$\frac{dx}{ds} = \cos\alpha = \frac{1}{\sqrt{1 + \left(\frac{dy}{dx}\right)^2}}, \quad \frac{dy}{ds} = \sin\alpha = \frac{1}{\sqrt{1 + \left(\frac{dx}{dy}\right)^2}},$$

$$ds = \sqrt{dx^2 + dy^2} = \sqrt{1 + \left(\frac{dy}{dx}\right)^2} \qquad dx = \sqrt{1 + \left(\frac{dx}{dy}\right)^2}\, dy.$$

$$ds = \sqrt{dr^2 + r^2 d\theta^2} = \sqrt{r^2 + \left(\frac{dr}{d\theta}\right)^2} \qquad d\theta = \sqrt{1 + r^2 \left(\frac{d\theta}{dr}\right)^2}\, dr.$$

If $x = r\cos\theta$, $y = r\sin\theta$,

$$dx = \cos\theta \cdot dr - r\sin\theta \cdot d\theta, \, dy = \sin\theta \cdot dr + r\cos\theta \cdot d\theta$$

The *radius of curvature* R at any point $P(x, y)$ of the curve $y = f(x)$ is

$$R = \frac{ds}{d\alpha} = \frac{\left[1 + \left(\frac{dy}{dx}\right)^2\right]^{\frac{3}{2}}}{\left(\frac{d^2y}{dx^2}\right)} = \frac{\left\{1 + \left[f'(x)\right]^2\right\}^{\frac{3}{2}}}{f''(x)},$$

$$R = \frac{\left[r^2 + \left(\frac{dr}{d\theta}\right)^2\right]^{\frac{3}{2}}}{r^2 + 2\left(\frac{dr}{d\theta}\right)^2 - r\frac{d^2r}{d\theta^2}}.$$

The *curvature* (K) at (x, y) is $K = \dfrac{1}{R}$.

The *center* of *curvature* corresponding to the point (x_1, y_1) on $y = f(x)$ is (h, k), where

$$h = x_1 - \frac{f'(x_1)\left\{1 + \left[f'(x_1)\right]^2\right\}}{f''(x_1)},$$

$$k = y_1 + \frac{1 + \left[f'(x_1)\right]^2}{f''(x_1)}.$$

115   MEAN ROLLE'S THEOREM

If $y = f(x)$ and its derivative $f'(x)$ be continuous on the interval (a, b), there exist a value of x somewhere between a and b such that

$$f(b) = f(a) + b - a)f'(x_1), a < x_1 < b.$$

Rolle's Theorem is a special case of the theorem of the mean with $f(a) = f(b) = 0$; that is, there exists at least one value of x between a and b for which $f'(x_1) = 0$.

116   EVALUATION OF INDETERMINATE FORMS

If $f(x)$ and $F(x)$ be two continuous functions of x having continuous derivatives, $f'(x)$ and $F'(x)$, then:

(a) If $\lim\limits_{x \to a} f(x) = 0$ and $\lim\limits_{x \to a} F(x) = 0$ and $\lim\limits_{x \to a} F'(x) \neq 0$,

[or if $\lim\limits_{x \to a} f(x) = \lim\limits_{x \to a} F(x) = \infty$], then

$$\lim_{x \to a} \frac{f(x)}{F(x)} = \lim_{x \to a} \frac{f'(x)}{F'(x)}.$$

(b) If $\lim\limits_{x \to a} f(x) = 0$ and $\lim\limits_{x \to a} F(x) = \infty$ [i.e., $F(x)$ becomes infinite as x approaches a as a limit], then $\lim\limits_{x \to a} [f(x).F(x)]$ may often be determined by writing

$$f(x).F(x) = \frac{f(x)}{1 / F(x)},$$

thus expressing the functions in the form of (a).

(c) If $\lim\limits_{x \to a} f(x) = \infty$ and $\lim\limits_{x \to a} F(x) = \infty$, then $\lim\limits_{x \to a} [f(x) - F(x)]$ may often be determined by writing

$$f(x) - F(x) = \frac{[1 / F(x)] - [1 / f(x)]}{1 / [f(x).F(x)]}$$

and using (a).

(d) The $\lim\limits_{x \to a} \left[f(x)^{F(x)} \right]$ may frequently be evaluated upon writing

$$f(x)^{F(x)} = e^{F(x).\log_e f(x)}$$

When one factor of the last exponent approaches zero and the other becomes infinite, the exponent is of the type considered in (b).

Thus, we are led to the indeterminate forms which are symbolized by

$$0 / 0, \, \infty / \infty, \, 0°, \, 1^\infty, \, \infty^0.$$

117   TAYLOR'S AND MACLAURIN'S THEOREM

Any function (continuous and having derivatives) may, in general, be expanded into a Taylor's series:

$$f(x) = f(a) + f'(a) \cdot \frac{(x-a)}{1!} + f''(a) \cdot \frac{(x-a)^2}{2!}$$

$$+ f'''(a) \cdot \frac{(x-a)^3}{3!} + \ldots + f^{(n-1)}(a) \cdot \frac{(x-a)^{n-1}}{(n-1)!} + R_n,$$

where a is any quantity for which $f(a), f'(a), f''(a)$, ... are finite.

If the series is to be used for approximating $f(x)$ for some value of x, then a should be picked so that the difference $(x - a)$ is numerically very small, and thus only a few terms of the series need to be used. The remainder, after n terms, is $R_n = f^{(n)}(x_1) \cdot (x - a)^n / n!$, where x_1 lies between a and x. R_n gives the limits of error in using n terms of the series for the approximation of the function:

$$n! = \lfloor n = 1 \cdot 2 \cdot 3 \cdot 4 \ldots n.$$

If $a = 0$, the above series is called Maclaurin's series:

$$f(x) = f(0) + f'(0)\frac{x}{1!} + f''(0)\frac{x^2}{2!} + f'''(0)\frac{x^3}{3!} + \ldots + f^{(n-1)}(0)\frac{x^{n-1}}{(n-1)!} + R_n.$$

118   SERIES

The following series may be obtained through the expansion of the functions by Taylor's or Maclaurin's Theorems. The expressions following a series indicate the region of convergence of the series, that is, the values of x for which R_n approaches zero as n becomes infinite, so that an approximation of the function may be obtained by using a number of terms of the series. If the region of convergence is not indicated, the series converges for all finite values of x. $(n! = 1 \cdot 2 \cdot 3 \ldots n)$. $\text{Log} e \, u \equiv \text{Log} \, u$.

(a) *Binomial Series*

$$(a + x)^n = a^n + na^{n-1}x + \frac{n(n-1)}{2!}a^{n-2}x^2$$

$$+ \frac{n(n-1)(n-2)}{3!}a^{n-3}x^3 + \ldots, x^2 < a^2.$$

If n is a positive integer, the series consists of $(n + 1)$ terms; otherwise, the number of terms is infinite.

$$(a - bx)^{-1} = \frac{1}{a}\left(1 + \frac{bx}{a} + \frac{b^2 x^2}{a^2} + \frac{b^3 x^3}{a^3} + \dots\right), \ b^2 x^2 < a^2.$$

(b) *Exponential, Logarithmic, and Trigonometric Series.*

$$e = 1 + \frac{1}{1!} + \frac{1}{2!} + \frac{1}{3!} + \frac{1}{4!} + \dots.$$

$$e^x = 1 + x + \frac{x^2}{2!} + \frac{x^3}{3!} + \frac{x^4}{4!} + \dots.$$

$$a^x = 1 + x\log a + \frac{(x\log a)^2}{2!} + \frac{(x\log a)^3}{3!} + \dots.$$

$$e^{-x^2} = 1 - x^2 + \frac{x^4}{2!} - \frac{x^6}{3!} + \frac{x^8}{4!} - \dots.$$

$$\log x = (x - 1) - \frac{1}{2}(x - 1)^2 + \frac{1}{3}(x - 1)^3 - \dots, \qquad 0 < x \leq 2.$$

$$\log x = \frac{x - 1}{x} + \frac{1}{2}\left(\frac{x - 1}{x}\right)^2 + \frac{1}{3}\left(\frac{x - 1}{x}\right)^3 + \dots, \qquad x > \frac{1}{2}.$$

$$\log x = 2\left[\frac{x - 1}{x + 1} + \frac{1}{3}\left(\frac{x - 1}{x + 1}\right)^3 + \frac{1}{5}\left(\frac{x - 1}{x + 1}\right)^5 + \dots\right], \qquad x > 0.$$

$$\log(1 + x) = x - \frac{x^2}{2} + \frac{x^3}{3} - \frac{x^4}{4} + \dots, \qquad -1 < x \leq 1.$$

$$\log(1 + x) = \log a + 2\left[\frac{x}{2a + x} + \frac{1}{3}\left(\frac{x}{2a + x}\right)^3 + \frac{1}{5}\left(\frac{x}{2a + x}\right)^5 + \dots\right],$$
$$a > 0, \ -a < x < +\infty.$$

$$\log\left(\frac{1 + x}{1 - x}\right) = 2\left(x + \frac{x^3}{3} + \frac{x^5}{5} + \frac{x^7}{7} + \dots\right), \qquad x^2 < 1.$$

$$\log\left(\frac{x+1}{x-1}\right) = 2\left[\frac{1}{x} + \frac{1}{3}\left(\frac{1}{x}\right)^3 + \frac{1}{5}\left(\frac{1}{x}\right)^5 + \frac{1}{7}\left(\frac{1}{x}\right)^7 + \cdots\right], \qquad x^2 > 1.$$

$$\log\left(\frac{x+1}{x}\right) = 2\left[\frac{1}{2x+1} + \frac{1}{3(2x+1)^3} + \frac{1}{5(2x+1)^5} + \cdots\right], \qquad x > 0.$$

$$\log\left(x + \sqrt{1+x^2}\right) = x - \frac{1}{2}\frac{x^3}{3} + \frac{1\cdot3}{2\cdot4}\frac{x^5}{5} - \frac{1\cdot3\cdot5}{2\cdot4\cdot6}\frac{x^7}{7} + \cdots, \qquad x^2 < 1.$$

$$\sin x = x - \frac{x^3}{3!} + \frac{x^5}{5!} - \frac{x^7}{7!} + \cdots.$$

$$\cos x = 1 - \frac{x^2}{2!} + \frac{x^4}{4!} - \frac{x^6}{6!} + \cdots$$

$$\tan x = x + \frac{x^3}{3} + \frac{2x^5}{15} + \frac{17x^7}{315} + \frac{62x^9}{2835} + \cdots, \qquad x^2 < \frac{\pi^2}{4}.$$

$$\sin^{-1} x = x + \frac{x^3}{6} + \frac{1}{2}\cdot\frac{3}{4}\cdot\frac{x^5}{5} + \frac{1}{2}\cdot\frac{3}{4}\cdot\frac{5}{6}\cdot\frac{x^7}{7} + \cdots, \qquad x^2 < 1.$$

$$\tan^{-1} x = x - \frac{1}{3}x^3 + \frac{1}{5}x^5 - \frac{1}{7}x^7 + \cdots, \qquad x^2 < 1.$$

$$= \frac{\pi}{2} - \frac{1}{x} + \frac{1}{3x^3} - \frac{1}{5x^5} + \cdots, \qquad x^2 > 1.$$

$$\log \sin x = \log x - \frac{x^2}{6} - \frac{x^4}{180} - \frac{x^6}{2835} - \cdots, \qquad x^2 < \pi^2.$$

$$\log \cos x = -\frac{x^2}{2} - \frac{x^4}{12} - \frac{x^6}{45} - \frac{17x^8}{2520} - \cdots, \qquad x^2 < \frac{\pi^2}{4}.$$

$$\log \tan x = \log x + \frac{x^2}{3} + \frac{7x^4}{90} + \frac{62x^6}{2835} + \cdots, \qquad x^2 < \frac{\pi^2}{4}.$$

$$e^{\sin x} = 1 + x + \frac{x^2}{2!} - \frac{3x^4}{4!} - \frac{8x^5}{5!} - \frac{3x^6}{6!} + \cdots$$

$$e^{\cos x} = e\left(1 - \frac{x^2}{2!} + \frac{4x^4}{4!} - \frac{31x^6}{6!} + \cdots\right).$$

$$e^{\tan x} = 1 + x + \frac{x^2}{2!} + \frac{3x^3}{3!} + \frac{9x^4}{4!} + \frac{37x^5}{5!} + \cdots, \qquad x^2 < \frac{\pi^2}{4}.$$

$$\sinh x = x + \frac{x^3}{3!} + \frac{x^5}{5!} + \frac{x^7}{7!} + \cdots.$$

$$\cosh x = 1 + \frac{x^2}{2!} + \frac{x^4}{4!} + \frac{x^6}{6!} + \cdots.$$

$$\tanh x = x - \frac{x^3}{3} + \frac{2x^5}{15} - \frac{17x^7}{315} + \cdots, \qquad x^2 < \frac{\pi^2}{4}.$$

$$\sinh^{-1} x = x - \frac{1}{2}\frac{x^3}{3} + \frac{1\cdot 3}{2\cdot 4}\frac{x^5}{5} - \frac{1\cdot 3\cdot 5}{2\cdot 4\cdot 6}\frac{x^7}{7} + \cdots, \qquad x^2 < 1.$$

$$\sinh^{-1} x = \log 2x + \frac{1}{2}\frac{1}{2x^2} - \frac{1\cdot 3}{2\cdot 4}\frac{1}{4x^4} + \frac{1\cdot 3\cdot 5}{2\cdot 4\cdot 6}\frac{1}{6x^6}\cdots, \qquad x > 1$$

$$\cosh^{-1} x = \log 2x - \frac{1}{2}\frac{1}{2x^2} - \frac{1\cdot 3}{2\cdot 4}\frac{1}{4x^4} - \frac{1\cdot 3\cdot 5}{2\cdot 4\cdot 6}\frac{1}{6x^6} - \cdots.$$

$$\tanh^{-1} x = x + \frac{x^3}{3} + \frac{x^5}{5} + \frac{x^7}{7} + \cdots, \qquad x^2 < 1.$$

119   PARTIAL DERIVATIVES: DIFFERENTIALS

If $z = f(x, y)$, is a function of two variables, then the derivative of z with respect to x, as x varies while y remains constant, is called the *first partial derivative of z with respect to x* and is denoted by $\frac{\partial z}{\partial x}$.

Similarly, the derivative of z with respect to y, as y varies while x remains constant, is called the *first partial derivative of z with respect to y* and is denoted by $\frac{\partial z}{\partial y}$.

Similarly, if $z = f(x, y, u, \ldots)$, then the first derivative of z with respect to x, as x varies while $y, u, \ldots$ remain constant, is called the first partial of z with respect to x and is denoted by $\dfrac{\partial z}{\partial x}$. Likewise, the second partial derivatives are defined as indicated below:

$$\frac{\partial^2 z}{\partial x^2} = \frac{\partial}{\partial x}\left(\frac{\partial z}{\partial x}\right); \frac{\partial^2 z}{\partial y^2} = \frac{\partial}{\partial y}\left(\frac{\partial z}{\partial y}\right); \frac{\partial^2 z}{\partial x\, \partial y} = \frac{\partial}{\partial x}\left(\frac{\partial z}{\partial y}\right) = \frac{\partial}{\partial y}\left(\frac{\partial z}{\partial x}\right) = \frac{\partial^2 z}{\partial y\, \partial x}.$$

If $z = f(x, y, ..., u)$, and $x, y, ..., u$ are functions of a single variable t, then

$$\frac{dz}{dt} = \frac{\partial z}{\partial x}\frac{dx}{dt} + \frac{\partial z}{\partial y}\frac{dy}{dt} + ... + \frac{\partial z}{\partial u}\frac{du}{dt},$$

$$dz = \frac{\partial z}{\partial x}dx + \frac{\partial z}{\partial y}dy + ... + \frac{\partial z}{\partial u}du.$$

If $F(x, y, z, ..., u) = 0$, then $\dfrac{\partial F}{\partial x}dx + \dfrac{\partial F}{\partial y}dy + ... + \dfrac{\partial F}{\partial u}du = 0$.

120 SURFACES: SPACE CURVES (SEE ANALYTIC GEOMETRY ART. 97-98)

The *tangent plane* to the surface $F(x, y, z) = 0$ at the point (x_1, y_1, z_1) on the surface is

$$(x - x_1)\left(\frac{\partial F}{\partial x}\right)_1 + (y - y_1)\left(\frac{\partial F}{\partial y}\right)_1 + (z - z_1)\left(\frac{\partial F}{\partial z}\right)_1 = 0,$$

where $\left(\dfrac{\partial F}{\partial x}\right)_1$ is the value of $\dfrac{\partial F}{\partial x}$ at (x_1, y_1, z_1), etc.

The equations of the *normal* to the surface at (x_1, y_1, z_1) are

$$\frac{x - x_1}{\left(\dfrac{\partial F}{\partial x}\right)_1} = \frac{y - y_1}{\left(\dfrac{\partial F}{\partial y}\right)_1} = \frac{z - z_1}{\left(\dfrac{\partial F}{\partial z}\right)_1}.$$

The *direction cosines* of the normal to the surface at the point (x_1, y_1, z_1) are proportional to

$$\left(\frac{\partial F}{\partial x}\right)_1, \left(\frac{\partial F}{\partial y}\right)_1, \left(\frac{\partial F}{\partial z}\right)_1.$$

Given the *space* curve $x = x(t)$, $y = y(t)$, $z = z(t)$. The direction cosines of the tangent line to the curve at any point are proportional to

$$\frac{dx}{dt}, \frac{dy}{dt}, \frac{dz}{dt}, \text{ or to } dx, dy, dz.$$

The equations of the *tangent line* to the curve at (x_1, y_1, z_1) on the curve are:

$$\frac{x - x_1}{\left(\dfrac{dx}{dt}\right)_1} = \frac{y - y_1}{\left(\dfrac{dy}{dt}\right)_1} = \frac{z - z_1}{\left(\dfrac{dz}{dt}\right)_1}$$

where $\left(\dfrac{dx}{dt}\right)_1$ is the value of $\dfrac{dx}{dt}$ at (x_1, y_1, z_1), etc.

INTEGRAL CALCULUS—I

121 INTRODUCTION

$F(x)$ is said to be an *indefinite integral* of $f(x)$, if the derivative of $F(x)$ is $f(x)$, or the differential of $F(x)$ is $f(x)\,dx$; symbolically:

$$F(x) = \int f(x)dx \text{ if } \frac{dF(x)}{dx} = f(x), \text{ or } dF(x) = f(x)dx.$$

In general: $\int f(x)dx = F(x) + C$, where C is an arbitrary constant.

122 FUNDAMENTAL THEOREMS ON INTEGRALS. SHORT TABLE OF INTEGRALS

(u and v denote functions of x; a, b *and* C denote constants).

1. $\int df(x) = f(x) + C.$

2. $d\int f(x)dx = f(x)dx.$

3. $\int 0 \cdot dx = C.$

4. $\int a\,f(x)dx = a\int f(x)dx.$

5. $\int (u \pm v)dx = \int u\,dx \pm \int v\,dx.$

6. $\int u\,dv = uv - \int v\,du.$

7. $\int \dfrac{u\,dv}{dx}dx = uv - \int v\dfrac{du}{dx}dx.$

8. $\int f(y)dx = \int \dfrac{(fy)dy}{\dfrac{dy}{dx}}.$

9. $\int u^n du = \dfrac{u^{n+1}}{n+1} + C, n \neq -1.$

10. $\int \dfrac{du}{u} = \log_e u + C,\ u > 0;\ \text{or},\ \log_e |u| + C, u \neq 0.$

11. $\int e^u du = e^u + C.$

12. $\int b^u du = \dfrac{b^u}{\log_e b} + C, b > 0, b \neq 1.$

13. $\int \sin u\,du = -\cos u + C.$

14. $\int \cos u\,du = \sin u + C.$

15. $\int \tan u\,du = \log_e \sec u + C = -\log_e \cos u + C.$

16. $\int \cot u\,du = \log_e \sin u + C = -\log_e \operatorname{cosec} u + C.$

17. $\int \sec u\,du = \log_e (\sec u + \tan u) + C = \log_e \tan\left(\dfrac{u}{2} + \dfrac{\pi}{4}\right) + C.$

18. $\int \operatorname{cosec} u\,du = \log_e (\operatorname{cosec} u - \cot u) + C = \log_e \tan\dfrac{u}{2} + C.$

19. $\int \sin^2 u\,du = \dfrac{1}{2}u - \dfrac{1}{2}\sin u \cos u + C.$

20. $\int \cos^2 u\,du = \dfrac{1}{2}u + \dfrac{1}{2}\sin u \cos u + C.$

21. $\int \sec^2 u\,du = \tan u + C.$

22. $\int \operatorname{cosec}^2 u \, du = -\cot u + C.$

23. $\int \tan^2 u \, du = \tan u - u + C.$

24. $\int \cot^2 u \, du = -\cot u - u + C.$

25. $\int \dfrac{du}{u^2 + a^2} = \dfrac{1}{a} \tan^{-1} \dfrac{u}{a} + C.$

26. $\int \dfrac{du}{u^2 - a^2} = \dfrac{1}{2a} \log_e \left(\dfrac{u-a}{u+a} \right) + C = -\dfrac{1}{a} \operatorname{ctnh}^{-1} \left(\dfrac{u}{a} \right) + C, \text{ if } u^2 + a^2,$

$\qquad = \dfrac{1}{2a} \log_e \left(\dfrac{a-u}{a+u} \right) + C = -\dfrac{1}{a} \tanh^1 \left(\dfrac{u}{a} \right) + C, \text{ if } u^2 + a^2.$

27. $\int \dfrac{du}{\sqrt{a^2 - u^2}} = \sin^{-1} \left(\dfrac{u}{a} \right) + C, a > 0; \text{ or, } \sin^{-1} \left(\dfrac{u}{|a|} \right) + C, |a| \neq 0.$

28. $\int \dfrac{du}{\sqrt{u^2 \pm a^2}} = \log_e \left(u + \sqrt{u^2 \pm a^2} \right)^* + C.$

29. $\int \dfrac{du}{\sqrt{2au - u^2}} = \cos^{-1} \left(\dfrac{a-u}{a} \right) + C.$

30. $\int \dfrac{du}{u\sqrt{u^2 - a^2}} = \dfrac{1}{a} \sec^{-1} \left(\dfrac{u}{a} \right) + C = \dfrac{1}{a} \cos^{-1} \dfrac{a}{u} + C.$

31. $\int \dfrac{du}{u\sqrt{a^2 \pm u^2}} = -\dfrac{1}{a} \log_e \left(\dfrac{a + \sqrt{a^2 \pm u^2}}{u} \right)^\circ + C.$

32. $\int \sqrt{a^2 - u^2} \cdot du = \dfrac{1}{2} \left(u\sqrt{a^2 - u^2} + a^2 \sin^{-1} \dfrac{u}{a} \right) + C.$

33. $\int \sqrt{u^2 \pm a^2} \, du = \dfrac{1}{2} \left[u\sqrt{u^2 \pm a^2} \pm a^2 \log_e \left(u + \sqrt{u^2 \pm a^2} \right) \right]^\circ + C.$

34. $\int \sinh u \, du = \cosh u + C.$

35. $\int \cosh u \, du = \sinh u + C.$

36. $\int \tanh u \, du = \log_e (\cosh u) + C.$

37. $\int \coth u \; du = \log_e (\sinh u) + C.$

38. $\int \text{sech} \; u \; du = \sin^{-1} (\tan u) + C.$

39. $\int \text{cosech} \; u \; du = \log_e \left(\tanh \dfrac{u}{2} \right) + C.$

40. $\int \text{sech} \; u \, . \, \tanh u \, . \, du = -\text{sech} \; u + C.$

41. $\int \text{cosech} \, . \, \coth u \, . \, du = -\text{cosech} \; u + C.$

* $\log_e \left(\dfrac{u + \sqrt{u^2 + a^2}}{a} \right) = \sinh^{-1} \left(\dfrac{u}{a} \right); \; \log_e \left(\dfrac{a + \sqrt{a^2 - u^2}}{u} \right) = \text{sech}^{-1} \left(\dfrac{u}{a} \right);$

* $\log_e \left(\dfrac{u + \sqrt{u^2 - a^2}}{a} \right) = \cosh^{-1} \left(\dfrac{u}{a} \right); \; \log_e \left(\dfrac{a + \sqrt{a^2 + u^2}}{u} \right) = \text{cosech}^{-1} \left(\dfrac{u}{a} \right).$

Definite Integrals

123 DEFINITION OF DEFINITE INTEGRAL

Let $f(x)$ be continuous for the interval from $x = a$ to $x = b$ inclusive. Divide this interval into n equal parts by the points $a, x_1, x_2, ..., x_{n-1}, b$ such that $\Delta x = (b - a)/n$. The definite integral of $f(x)$ with respect to x between the limits $x = a$ to $x = b$ is:

$$\int_a^b f(x)dx = \lim_{n \to \infty} \left[f(a)\Delta x + f(x_1)\Delta x + f(x_2)\Delta x + ... + f(x_{n-1})\Delta x \right].$$

$$\int_a^b f(x)dx = \left[\int f(x)dx \right]_a^b = \left[F(x) \right]_a^b = F(b) - F(a),$$

where $F(x)$ is a function whose derivative with respect to x is $f(x)$.

124 APPROXIMATE VALUES OF DEFINITE INTEGRAL

Approximate values of the above definite integral are given by the rules of § 41, where $y_0, y_1, y_2 ..., y_{n-1}, y_n$ are the values of $f(x)$ for $x = a, x_1, x_2, ..., x_{n-1}, b$, respectively, and $h = (b - a)/n$.

125 SOME FUNDAMENTAL THEOREMS

$$\int_a^b \left[f_1(x) + f_2(x) + \dots + f_n(x) \right] dx = \int_a^b f_1(x) dx + \int_a^b f_2(x) dx + \dots + \int_a^b f_n(x) dx.$$

$$\int_a^b k f(x) dx = k \int_a^b f(x) dx, \text{ if } k \text{ is a constant.}$$

$$\int_a^b f(x) dx = -\int_b^a f(x) dx.$$

$$\int_a^b f(x) dx = \int_a^c f(x) dx + \int_c^b f(x) dx.$$

$$\int_a^b f(x) dx = (b - a) f(x_1), \text{ where } x_1 \text{ lies between } a \text{ and } b.$$

$$\int_a^\infty f(x) dx = \lim_{t \to \infty} \int_a^t f(x) dx.$$

$$\int_{-\infty}^b f(x) dx = \lim_{t \to \infty} \int_{-t}^b f(x) dx.$$

$$\int_{-\infty}^{+\infty} f(x) dx = \int_{-\infty}^c f(x) dx + \int_c^\infty f(x) dx.$$

If $f(x)$ has a singular point* at $x = b$, $b \neq a$,

$$\int_a^b f(x) dx = \lim_{e \to 0} \int_a^{b-e} f(x) dx.$$

The mean value of the function $f(x)$ on the interval (a, b) is:

$$\frac{1}{b-a} \int_a^b f(x) dx.$$

Some Applications of the Definite Integral

126 PLANE AREA

The area bounded by $y = f(x)$, $y = 0$, $x = a$, $x = b$, where y has the same sign for all values of x between a and b, is:

$$A = \int_a^b f(x) dx, dA = f(x) dx.$$

The area bounded by the curve $r = f(\theta)$ and the two radii $\theta = \alpha$, $\theta = \beta$, is

$$A = \frac{1}{2}\int_{\alpha}^{\beta}[f(\theta)]^2\, d\theta, \quad dA = \frac{1}{2}r^2 \cdot d\theta.$$

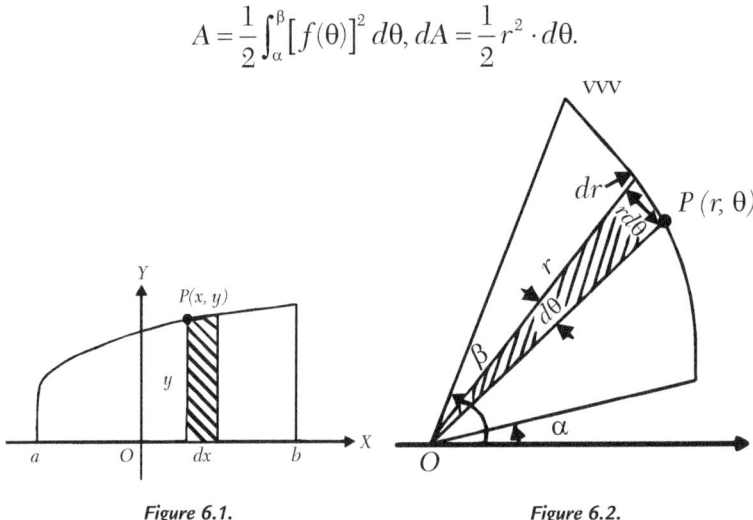

Figure 6.1.   Figure 6.2.

127 LENGTH OF ARC

The length (s) of the arc of a plane curve $f(x, y) = 0$ from the point (a, c) to the point (b, d) is

$$s = \int_a^b \sqrt{1 + \left(\frac{dy}{dx}\right)^2} \cdot dx = \int_c^d \sqrt{1 + \left(\frac{dx}{dy}\right)^2} \cdot dy.$$

If the equation of the curve is $x = f(t), y = f(t)$, the length of the arc from $t = a$ to $t = b$ is

$$s = \int_a^b \sqrt{\left(\frac{dx}{dt}\right)^2 + \left(\frac{dy}{dt}\right)^2} \cdot dt.$$

If the equation of the curve is $r = f(\theta)$, then

$$s = \int_{\theta_1}^{\theta_2} \sqrt{r^2 + \left(\frac{dr}{d\theta}\right)^2} \cdot d\theta = \int_{r_1}^{r_2} \sqrt{r^2 \left(\frac{d\theta}{dr}\right)^2 + 1} \cdot dr.$$

128 VOLUME BY PARALLEL SECTIONS

If the plane perpendicular to the x-axis at $(x, 0, 0)$ cuts from a given solid a section whose area is $A(x)$, then the volume of that part of the solid between $x = a$ and $x = b$ is:

$$\int_a^b A(x)dx.$$

129 VOLUME OF REVOLUTION

The volume of a solid of revolution generated by revolving that portion of the curve $y = f(x)$ between $x = a$ and $x = b$

(a) about the x-axis is $\pi\int_a^b y^2 dx$;

(b) about the y-axis is $\pi\int_c^d x^2 dy$, where c and d are the values of y corresponding to the values a and b of x.

130 AREA OF SURFACE OF REVOLUTION

The area of the surface of a solid of revolution generated by revolving the curve $y = f(x)$ between $x = a$ and $x = b$

(a) about the x-axis is $2\pi\int_a^b y\sqrt{1+\left(\dfrac{dy}{dx}\right)^2}\cdot dx,$

(b) about the y = axis is $2\pi\int_c^d x\sqrt{1+\left(\dfrac{dx}{dy}\right)^2}\cdot dy.$

131 PLANE AREAS BY DOUBLE INTEGRATION

(a) Rectangular coordinates

$$A = \int_a^b \int_{\phi(x)}^{f(x)} dy\,dx \text{ or } \int_c^d \int_{\xi(y)}^{\psi(y)} dx\,dy;$$

(b) Polar coordinates

$$A = \int_{\theta_1}^{\theta_2} \int_{f_1(\theta)}^{f_2(\theta)} r\,dr\,d\theta \text{ or } \int_{r_1}^{r_2} \int_{\phi_1(r)}^{\phi_2(r)} r\,d\theta\,dr.$$

132 VOLUMES BY DOUBLE INTEGRATION

If $z = f(x, y)$,

$$V = \int_a^b \int_{\phi(x)}^{\psi(x)} f(x, y) \, dy \, dx \text{ or } \int_c^d \int_{\alpha(y)}^{\beta(y)} f(x, y) \, dx \, dy.$$

133 VOLUMES BY TRIPLE INTEGRATION

(a) Rectangular coordinates

$$V = \iiint dx \, dy \, dz;$$

(b) Cylindrical Coordinates

$$V = \iiint r \, dr \, d\theta \, dz;$$

(c) Spherical Coordinates

$$V = \iiint \rho^2 \sin\phi \, d\theta \, d\phi \, d\rho.,$$

where the limits of integration must be supplied. Other formulas may be obtained by changing the order of integration.

Area of Surface $z = f(x, y)$

$$A = \iint \sqrt{\left(\frac{\partial z}{\partial x}\right)^2 + \left(\frac{\partial z}{\partial y}\right)^2 + 1} \cdot dy \, dx,$$

where the limits of integration must be supplied.

134 MASS

The mass of a body of density δ is

$$m = \int dm, \, dm = \delta \cdot dA, \text{ or } \delta \cdot ds, \text{ or } \delta \cdot dV, \text{ or } \delta \cdot dS,$$

where dA, ds, dV, and ds are, respectively, the elements of area, length, volume, and surface of Art. 125 to Art. 132.

135 DENSITY

If δ is a variable (or constant) density (mass per unit of element), and $\bar{\delta}$ is the mean density of a solid of volume V, then

$$\bar{\delta} = \frac{\int \delta \, dV}{\int dV}.$$

136 MOMENT

The moments M_{yz}, M_{xz}, and M_{xy}, of a mass m with respect to the coordinate planes (as indicated by the subscripts) are:

$$M_{yz} = \int x \, dm, \, M_{xz} = \int y \, dm, \, M_{xy} = \int z \, dm.$$

137 CENTROID OF MASS OR CENTER OF GRAVITY

The coordinates $\left(\bar{x}, \bar{y}, \bar{z}\right)$ of the centroid of a mass m are:

$$\bar{x} = \frac{\int x \, dm}{\int dm}, \quad \bar{y} = \frac{\int y \, dm}{\int dm}, \quad \bar{z} = \frac{\int z \, dm}{\int dm}.$$

NOTE *In the above equations x, y, z are the coordinates of the center of gravity of the element dm.*

138 CENTROID OF SEVERAL MASSES

The x-coordinate $(\bar{x})$ of the centroid of several masses $m_1, m_2, ..., m_n$, having $\bar{x}_1, \bar{x}_2 \ldots , \bar{x}_n$, respectively, as the x-coordinates of their centroids, is:

$$\bar{x} = \frac{m_1\bar{x}_1 + m_2\bar{x}_2 + ... + m_n\bar{x}_n}{m_1 + m_2 + ... + m_n}.$$

Similar formulas hold for the other coordinates $\bar{y}$, $\bar{z}$.

139(A)   MOMENT OF INERTIA (SECOND MOMENT)

The moments of inertia (I).

(a) for a plane curve about the x-axis, y-axis, and origin, respectively, are:

$$I_x = \int y^2 \, ds, \qquad I_y = \int x^2 \, ds, \qquad I_0 = \int (x^2 + y^2) \, ds;$$

(b) for a plane area about the x-axis, y-axis, and origin, respectively, are:

$$I_x = \int y^2 \, dA, \qquad I_y = \int x^2 \, dA, \qquad I_0 = \int (x^2 + y^2) \, dA;$$

(c) for a solid of mass m about the yz, xz, and xy-planes, x-axis, etc., respectively, are:

$$I_{yz} = \int y^2 \, dm, \qquad I_{xz} = \int y^2 \, dm, \qquad I_{xy} = \int z^2 \, dm,$$

$$I_x = I_{xz} + I_{xy}, \text{ etc.}$$

The limits of integration are to be supplied.

139(B) THEOREM OF PARALLEL AXES

Let L be any line in space and L_g a line parallel to L, passing through the centroid of the body of mass m. If d is the distance between the lines L and L_g, then

$$I_L = I_{Lg} + d^2 m,$$

where I_L and I_{Lg} are the moments of inertia of the body about the lines L and L_g, respectively.

140 RADIUS OF GYRATION

If I is the moment of inertia of a mass m, and K is the radius of gyration, $I = mK^2$.

Similarly for areas, lengths, volumes, etc.

If masses (or areas, etc.) m_1, m_2, ..., m_n, have, respectively, the radii of gyration k_1 k_2, ..., k_n, with respect to a line or plane, then with respect to this line or plane, the several masses taken together have the radius of gyration K,

where
$$K^2 = \frac{m_1 k_1^2 + m_2 k_2^2 + ... m_n k_n^2}{m_1 + m_2 + ... + m_n}.$$

141 WORK

The work W done in moving a particle from $s = a$ to $s = b$ by a force whose component expressed as function of s in the direction of motion is F_s, is

$$W = \int_{s=a}^{s=b} F_s \, ds, \quad dW = F_s ds.$$

142 PRESSURE

The pressure (P) against an area vertical to the surface of a liquid and between the depths a and b is:

$$P = \int_{y=a}^{y=b} wly \, dy, \quad dp = wly \, dy,$$

where w is the weight of liquid per unit volume, y is the depth beneath surface of liquid of a horizontal element of area, and l is the length of the horizontal element of area expressed in terms of y.

143 CENTER OF PRESSURE

The depth $\bar{y}$ of the center of pressure against an area vertical to the surface of the liquid and between the depths a and b is:

$$\bar{y} = \frac{\int_{y=a}^{y=b} y dp}{\int_{y=a}^{y=b} dp}.$$

INTEGRAL CALCULUS—II

Certain Elementary Processes

144 TO INTEGRATE $\int R(x) \cdot dx$, WHERE $R(X)$ IS RATIONAL FUNCTION OF X.

Write $R(x)$ in the form of a fraction whose terms are polynomials. If the fraction is improper, (*i.e.*, the degree of the denominator is less than or equal to the degree of the numerator), divide the numerator by the denominator and thus write $R(x)$ as the sum of a quotient $Q(x)$ and a proper fraction $P(x)$. (In a proper fraction the degree of the numerator is less than the degree of the denominator.) The polynomial $Q(x)$ is readily integrated. To integrate $P(x)$ separate it into a sum of partial fractions (as indicated below), and integrate each term of the sum separately.

To separate the proper fraction $P(x)$ into partial fractions, write $P(x) = f(x)/\psi(x)$, where $f(x)$ and $\psi(x)$ are polynomials,

$$\psi(x) = (x-a)^p (x-b)^q (x-c)^r \dots,$$

and the constants a, b, and c, are all different. By algebra, there exist constants $A_1, A_2, \dots, B_1, B_2, \dots$, such that

$$\frac{f(x)}{\psi(x)} = \frac{A_1}{(x-a)} + \frac{A_2}{(x-a)^2} + \dots + \frac{A_p}{(x-a)^p} + \frac{B_1}{(x-b)}$$

$$+ \frac{B_2}{(x-b)^2} + \dots + \frac{B_q}{(x-b)^q} + \dots$$

The separate terms of this sum may be integrated by the formulas

$$\int \frac{dx}{(x-\alpha)^t} = \frac{-1}{(t-1)(x-\alpha)^{t-1}}, t > 1, \int \frac{dx}{(x-\alpha)} = \log(x-\alpha).$$

If $f(x)$ and $\psi(x)$ have real coefficients and $\psi(x) = 0$ has imaginary roots the above method leads to imaginary quantities. To avoid this, separate $P(x)$ in a different way. As before, corresponding to each p-fold real root a of $\psi(x)$, use the sum

$$\frac{A_1}{(x-a)} + \frac{A_2}{(x-a)^2} + \ldots + \frac{A_p}{(x-a)^p}.$$

To each λ-fold real quadratic factor $x^2 + \alpha x + \beta$, of $\psi(x)$ which does not factor into real linear factors, use, instead of the two sets of terms occurring in the first expansion of $R(x)$ and dependent on the conjugate complex roots of $x^2 + \alpha x + \beta = 0$, sums of the forms

$$\frac{D_1 x + E_1}{\left(x^2 + \alpha x + \beta\right)} + \frac{D_2 x + E_2}{\left(x^2 + \alpha x + \beta\right)^2} + \ldots + \frac{D_\lambda x + E_\lambda}{\left(x^2 + \alpha x + \beta\right)^\lambda},$$

where the quantities $D_1, D_2, \ldots, E_1, E_2, \ldots$, are real constants.

These new sums may be separately integrated by means of Integral formulas 149 to 156, etc.

145 TO INTEGRATE AN IRRATIONAL ALGEBRAIC FUNCTION

If no convenient method of integration is apparent the integration may frequently be performed by means of a change of variable. For example, if R is a rational function of two arguments and n is an integer, then to integrate

(a) $\int R\left[x, (ax+b)^{\frac{1}{n}}\right] dx$, let $(ax + b) = y^n$, whence the integral reduces to $\int P(y)\, dy$, where $P(y)$ is a rational function of y and may be integrated as in Art. 143;

(b) $\int R\left[x,\left(x^2+bx+c\right)^{\frac{1}{2}}\right]dx,$ let $\left(x^2+bx+c\right)^{\frac{1}{2}}=z-x,$ reduce

R to a rational function of z, and proceed as in Art. 143;

(c) $\int R(\sin x,\cos x)dx,$ let $\tan\dfrac{x}{2}=t,$ whence

$$\sin x=\frac{2t}{1+t^2},\ \cos x=\frac{1-t^2}{1+t^2},\ dx=\frac{2dt}{1+t^2},$$

reduce to a rational function of t and proceed as in Art. 143.

146 TO INTEGRATE EXPRESSIONS CONTAINING $\sqrt{a^2-x^2},\ \sqrt{x^2\pm a^2}$

Expressions containing these radicals can frequently be integrated after making the following transformations:

(a) if $\sqrt{a^2-x^2}$ occurs, let $x=a\sin t$;

(b) if $\sqrt{x^2-a^2}$ occurs, let $x=a\sec t$;

(c) if $\sqrt{x^2+a^2}$ occurs, let $x=a\tan t$;

147 TO INTEGRATE EXPRESSIONS CONTAINING TRIGONOMETRIC FUNCTIONS

The integration of such functions may frequently be facilitated by the use of the identities of Art. 58.

148 INTEGRATION BY PARTS

The relation

$$\int u\,dv=u\,v-\int v\,du$$

is often effective in reducing a given integral to one or more simpler integrals.

149 TABLE OF INTEGRALS

In the following table, the constant of integration, C, is omitted but should be added to the result of every integration. The letter x represents any variable; u represents any function of x; the remaining letters represent arbitrary constants, unless otherwise indicated; all angles are in radians. **Unless otherwise mentioned $\log_e u \equiv \log u$.**

Expressions Containing (ax + b)

42. $\int (ax+b)^n \, dx = \dfrac{1}{a(n+1)} (ax+b)^{n+1}, \, n \neq -1.$

43. $\int \dfrac{dx}{ax+b} = \dfrac{1}{a} \log_e (ax+b).$

44. $\int \dfrac{dx}{(ax+b)^2} = -\dfrac{1}{a(ax+b)}.$

45. $\int \dfrac{dx}{(ax+b)^3} = -\dfrac{1}{2a(ax+b)^2}.$

46. $\int x(ax+b)^n \, dx = \dfrac{1}{a^2(n+2)} (ax+b)^{n+2}$
$$-\dfrac{b}{a^2(n+1)} (ax+b)^{n+1}, \, n \neq -1, -2.$$

47. $\int \dfrac{xdx}{ax+b} = \dfrac{x}{a} - \dfrac{b}{a^2} \log(ax+b).$

48. $\int \dfrac{xdx}{(ax+b)^2} = \dfrac{b}{a^2(ax+b)} + \dfrac{1}{a^2} \log(ax+b).$

49. $\int \dfrac{xdx}{(ax+b)^3} = \dfrac{b}{2a^2(ax+b)^2} - \dfrac{1}{a^2(ax+b)}.$

50. $\int x^2(ax+b)^n \, dx = \dfrac{1}{a^3} \left[\dfrac{(ax+b)^{n+3}}{n+3} - 2b \dfrac{(ax+b)^{n+2}}{n+2} \right.$
$$\left. +b^2 \dfrac{(ax+b)^{n+1}}{n+1} \right], \, n \neq -1, -2, -3.$$

51. $\int \dfrac{x^2 dx}{ax+b} = \dfrac{1}{a^3}\left[\dfrac{1}{2}(ax+b)^2 - 2b(ax+b) + b^2 \log(ax+b)\right].$

52. $\int \dfrac{x^2 dx}{(ax+b)^2} = \dfrac{1}{a^3}\left[(ax+b) - 2b \log(ax+b) - \dfrac{b^2}{ax+b}\right].$

53. $\int \dfrac{x^2 dx}{(ax+b)^3} = \dfrac{1}{a^3}\left[\log(ax+b) + \dfrac{2b}{ax+b} - \dfrac{b^2}{2(ax+b)^2}\right].$

54. $\int x^m (ax+b)^n\, dx = \dfrac{1}{a(m+n+1)}$
$$\left[x^m (ax+b)^{n+1} - mb\int x^{m-1}(ax+b)^n\, dx\right],$$
$$= \dfrac{1}{m+n+1}\left[x^{m+1}(ax+b)^n + nb\int x^m (ax+b)^{n-1}\, dx\right],$$
$$m > 0,\ m+n+1 \neq 0.$$

55. $\int \dfrac{dx}{x(ax+b)} = \dfrac{1}{b}\log\dfrac{x}{ax+b}.$

56. $\int \dfrac{dx}{x^2(ax+b)} = -\dfrac{1}{bx} + \dfrac{a}{b^2}\log\dfrac{ax+b}{x}.$

57. $\int \dfrac{dx}{x^3(ax+b)} = \dfrac{2ax-b}{2b^2 x^2} + \dfrac{a^2}{b^3}\log\dfrac{x}{ax+b}.$

58. $\int \dfrac{dx}{x(ax+b)^2} = \dfrac{1}{b(ax+b)} - \dfrac{1}{b^2}\log\dfrac{ax+b}{x}.$

59. $\int \dfrac{dx}{x(ax+b)^3} = \dfrac{1}{b^3}\left[\dfrac{1}{2}\left(\dfrac{ax+2b}{ax+b}\right)^2 + \log\dfrac{x}{ax+b}\right].$

60. $\int \dfrac{dx}{x^2(ax+b)^2} = -\dfrac{b+2ax}{b^2 x(ax+b)} + \dfrac{2a}{b^3}\log\dfrac{ax+b}{x}.$

61. $\int \sqrt{ax+b}\ dx = \dfrac{2}{3a}\sqrt{(ax+b)^3}$.

62. $\int x\sqrt{ax+b}\ dx = \dfrac{2(3ax-2b)}{15a^2}\sqrt{(ax+b)^3}$.

63. $\int x^2\sqrt{ax+b}\ dx = \dfrac{2\left(15a^2x^2-12abx+8b^2\right)\sqrt{(ax+b)^3}}{105a^3}$.

64. $\int x^3\sqrt{ax+b}\ dx = \dfrac{2\left(35a^3x^3-30a^2bx^2+24ab^2x-16b^3\right)\sqrt{(ax+b)^3}}{315a^4}$.

65. $\int x^n\sqrt{ax+b}\ dx = \dfrac{2}{a^{n+1}}\int u^2\left(u^2-b\right)^n du,\ u=\sqrt{ax+b}$.

66. $\int \dfrac{\sqrt{ax+b}}{x}\ dx = 2\sqrt{ax+b}+b\int\dfrac{dx}{x\sqrt{ax+b}}$.

67. $\int \dfrac{dx}{\sqrt{ax+b}} = \dfrac{2\sqrt{ax+b}}{a}$.

68. $\int \dfrac{xdx}{\sqrt{ax+b}} = \dfrac{2(ax-2b)}{3a^2}\sqrt{ax+b}$.

69. $\int \dfrac{x^2 dx}{\sqrt{ax+b}} = \dfrac{2\left(3a^2x^2-4abx+8b^2\right)}{15a^3}\sqrt{ax+b}$.

70. $\int \dfrac{x^3 dx}{\sqrt{ax+b}} = \dfrac{2\left(5a^3x^3-6a^2bx^2+8ab^2x-16b^3\right)}{35a^4}\sqrt{ax+b}$.

71. $\int \dfrac{x^n dx}{\sqrt{ax+b}} = \dfrac{2}{a^{n+1}}\int\left(u^2-b\right)^n du,\quad u=\sqrt{ax+b}$.

72. $\int \dfrac{dx}{x\sqrt{ax+b}} = \dfrac{1}{\sqrt{b}}\log\dfrac{\sqrt{ax+b}-\sqrt{b}}{\sqrt{ax+b}+\sqrt{b}},\quad$ for $b>0$.

73. $\int \dfrac{dx}{x\sqrt{ax+b}} = \dfrac{2}{\sqrt{-b}} \tan^{-1} \sqrt{\dfrac{ax+b}{-b}}, b<0;$

$$\dfrac{-2}{\sqrt{b}} \tan h^{-1} \sqrt{\dfrac{ax+b}{b}}, b>0.$$

74. $\int \dfrac{dx}{x^2 \sqrt{ax+b}} = -\dfrac{\sqrt{ax+b}}{bx} - \dfrac{a}{2b} \int \dfrac{dx}{x\sqrt{ax+b}}.$

75. $\int \dfrac{dx}{x^3 \sqrt{ax+b}} = \dfrac{\sqrt{ax+b}}{2bx^2} + \dfrac{3a\sqrt{ax+b}}{4b^2 x} + \dfrac{3a^2}{8b^2} \int \dfrac{dx}{x\sqrt{ax+b}}.$

76. $\int \dfrac{dx}{x^n (ax+b)^m} = -\dfrac{1}{b^{m+n-1}} \int \dfrac{(u-a)^{m+n-2} \, du}{u^m}, u = \dfrac{ax+b}{x}.$

77. $\int (ax+b)^{\pm \frac{n}{2}} \, dx = \dfrac{2(ax+b)^{\frac{2\pm n}{2}}}{a(2 \pm n)}.$

78. $\int x(ax+b)^{\pm \frac{n}{2}} \, dx = \dfrac{2}{a^2} \left[\dfrac{(ax+b)^{\frac{4\pm n}{2}}}{4 \pm n} - \dfrac{b(ax+b)^{\frac{2\pm n}{2}}}{2 \pm n} \right].$

79. $\int \dfrac{dx}{x(ax+b)^{\frac{n}{2}}} = \dfrac{1}{b} \int \dfrac{dx}{x(ax+b)^{\frac{n-2}{2}}} - \dfrac{a}{b} \int \dfrac{dx}{(ax+b)^{\frac{n}{2}}}.$

80. $\int \dfrac{x^m \, dx}{\sqrt{ax+b}} = \dfrac{2x^m \sqrt{ax+b}}{(2m+1)a} - \dfrac{2mb}{(2m+1)a} \int \dfrac{x^{m-1} \, dx}{\sqrt{ax+b}}.$

81. $\int \dfrac{dx}{x^n \sqrt{ax+b}} = \dfrac{-\sqrt{ax+b}}{(n-1)bx^{n-1}} - \dfrac{(2n-3)a}{(2n-2)b} \int \dfrac{dx}{x^{n-1} \sqrt{ax+b}}.$

82. $\int \dfrac{(ax+b)^{\frac{n}{2}}}{x} \, dx = a \int (ax+b)^{\frac{n-2}{2}} \, dx + b \int \dfrac{(ax+b)^{\frac{n-2}{2}}}{x} \, dx.$

83. $\int \dfrac{dx}{(ax+b)(cx+d)} = \dfrac{1}{bc-ad} \log \dfrac{cx+d}{ax+b}, bc - ad \neq 0.$

84. $\int \dfrac{dx}{(ax+b)^2(cx+d)} = \dfrac{1}{bc-ad}\left[\dfrac{1}{ax+b} + \dfrac{c}{bc-ad}\log\left(\dfrac{cx+d}{ax+b}\right)\right],$

$$bc-ad \neq 0.$$

85. $\int (ax+b)^n (cx+d)^m \, dx$

$$= \dfrac{1}{(m+n+1)a}\left[(ax+b)^{n+1}(cx+d)^m - m(bc-ad)\right.$$

$$\left. \int (ax+b)^n (cx+d)^{m-1} \, dx\right].$$

86. $\int \dfrac{dx}{(ax+b)^n (cx+d)^m}$

$$= \dfrac{-1}{(m-1)(bc-ad)}\left[\dfrac{1}{(ax+b)^{n-1}(cx+d)^{m-1}} + a(m+n-2)\right.$$

$$\left. \int \dfrac{dx}{(ax+b)^n (cx+d)^{m-1}}\right],$$

$$m > 1, n > 0, bc-ad \neq 0.$$

87. $\int \dfrac{(ax+b)^n}{(cx+d)^m} \, dx$

$$= -\dfrac{1}{(m-1)(bc-ad)}\left[\dfrac{(ax+b)^{n+1}}{(cx+d)^{m-1}} + (m-n-2)a\int \dfrac{(ax+b)^n \, dx}{(cx+d)^{m-1}}\right],$$

$$= \dfrac{-1}{(m-n-1)c}\left[\dfrac{(ax+b)^n}{(cx+d)^{m-1}} + n(bc-ad)\int \dfrac{(ax+b)^{n-1}}{(cx+d)^m} \, dx\right].$$

88. $\int \dfrac{xdx}{(ax+b)(cx+d)} = \dfrac{1}{bc-ad}\left[\dfrac{b}{a}\log(ax+b) - \dfrac{d}{c}\log(cx+d)\right],$

$$bc-ad \neq 0.$$

89. $\int \dfrac{xdx}{(ax+b)^2(cx+d)} = \dfrac{1}{bc-ad}$

$$\left[-\dfrac{b}{a(ax+b)} - \dfrac{d}{bc-ad}\log\dfrac{cx+d}{ax+b}\right], bc-ad \neq 0.$$

90. $\int \dfrac{cx+d}{\sqrt{ax+b}} \, dx = \dfrac{2}{3a^2}(3ad - 2bc + acx)\sqrt{ax+b}.$

91. $\int \dfrac{\sqrt{ax+b}}{cx+d}\,dx = \dfrac{2\sqrt{ax+b}}{c} - \dfrac{2}{c}\sqrt{\dfrac{ad-bc}{c}}\ \tan^{-1}$

$\sqrt{\dfrac{c(ax+b)}{ad-bc}}, c>0, ad>bc.$

92. $\int \dfrac{\sqrt{ax+b}}{cx+d}\,dx = \dfrac{2\sqrt{ax+b}}{c} + \dfrac{1}{c}\sqrt{\dfrac{bc-ad}{c}}$

$\log \dfrac{\sqrt{c(ax+b)} - \sqrt{bc-ad}}{\sqrt{c(ax+b)} + \sqrt{bc-ad}}, c>0, bc>ad.$

93. $\int \dfrac{dx}{(cx+d)\sqrt{ax+b}} = \dfrac{2}{\sqrt{c}\sqrt{ad-bc}}\ \tan^{-1}\sqrt{\dfrac{c(ax+b)}{ad-bc}}\ \ c>0, ad>bc.$

94. $\int \dfrac{dx}{(cx+d)\sqrt{ax+b}} = \dfrac{1}{\sqrt{c}\sqrt{bc-ad}}\ \log \dfrac{\sqrt{c(ax+b)} - \sqrt{bc-ad}}{\sqrt{c(ax+b)} + \sqrt{bc-ad}},$

$c>0, bc>ad.$

Expressions Containing $ax^2+c,\ ax^n+c,\ x^2 \pm p^2,$ **and** $p^2 - x^2.$

95. $\int \dfrac{dx}{p^2 - x^2} = \dfrac{1}{p}\tan^{-1}\dfrac{x}{p},\ \text{or} -\dfrac{1}{p}\operatorname{ctn}^{-1}\left(\dfrac{x}{p}\right).$

96. $\int \dfrac{dx}{p^2 - x^2} = \dfrac{1}{2p}\log\dfrac{p+x}{p-x},\ \text{or}\ \dfrac{1}{p}\tanh^{-1}\left(\dfrac{x}{p}\right).$

97. $\int \dfrac{dx}{ax^2 + c} = \dfrac{1}{\sqrt{ac}}\tanh^{-1}\left(x\sqrt{\dfrac{a}{c}}\right),\ a \text{ and } c>0.$

98. $\int \dfrac{dx}{ax^2 + c} = \dfrac{1}{2\sqrt{-ac}}\log\dfrac{x\sqrt{a} - \sqrt{-c}}{x\sqrt{a} + \sqrt{-c}}, a>0, c<0.$

$= \dfrac{1}{2\sqrt{-ac}}\log\dfrac{\sqrt{c} + x\sqrt{-a}}{\sqrt{c} - x\sqrt{-a}}, a<0, c>0.$

99. $\int \dfrac{dx}{\left(ax^2 + c\right)^n} = \dfrac{1}{2(n-1)c} \cdot \dfrac{x}{\left(ax^2 + c\right)^{n-1}}$

$$+ \dfrac{2n-3}{2(n-1)c} \int \dfrac{dx}{\left(ax^2 + c\right)^{n-1}}, n > 1.$$

100. $\int x\left(ax^2 + c\right)^n dx = \dfrac{1}{2a} \dfrac{\left(ax^2 + c\right)^{n+1}}{n+1}, n \neq 1.$

101. $\int \dfrac{x}{ax^2 + c} dx = \dfrac{1}{2a} \log\left(ax^2 + c\right).$

102. $\int \dfrac{dx}{x\left(ax^2 + c\right)} = \dfrac{1}{2c} \log \dfrac{x^2}{ax^2 + c}.$

103. $\int \dfrac{dx}{x^2\left(ax^2 + c\right)} = -\dfrac{1}{cx} - \dfrac{a}{c} \int \dfrac{dx}{ax^2 + c}.$

104. $\int \dfrac{x^2 dx}{ax^2 + c} = \dfrac{x}{a} - \dfrac{c}{a} \int \dfrac{dx}{ax^2 + c}.$

105. $\int \dfrac{x^n dx}{ax^2 + c} = \dfrac{x^{n-1}}{a(n-1)} - \dfrac{c}{a} \int \dfrac{x^{n-1} dx}{ax^2 + c}, n \neq 1.$

106. $\int \dfrac{x^2 dx}{\left(ax^2 + c\right)^n} = -\dfrac{1}{2(n-1)a} \cdot \dfrac{x}{\left(ax^2 + c\right)^{n-1}} + \dfrac{1}{2(n-1)a} \int \dfrac{dx}{\left(ax^2 + c\right)^{n-1}}.$

107. $\int \dfrac{dx}{x^2\left(ax^2 + c\right)^n} = \dfrac{1}{c} \int \dfrac{dx}{x^2\left(ax^2 + c\right)^{n-1}} - \dfrac{a}{c} \int \dfrac{dx}{\left(ax^2 + c\right)^n}.$

108. $\int \sqrt{x^2 \pm p^2}\, dx = \dfrac{1}{2}\left[x\sqrt{x^2 \pm p^2} \pm p^2 \log\left(x + \sqrt{x^2 \pm p^2}\right)\right].$

109. $\int \sqrt{p^2 - x^2}\, dx = \dfrac{1}{2}\left[x\sqrt{p^2 - x^2} + p^2 \sin^{-1}\left(\dfrac{x}{p}\right)\right].$

110. $\int \dfrac{dx}{\sqrt{x^2 \pm p^2}} = \log\left(x + \sqrt{x^2 \pm p^2}\right).$

111. $\int \dfrac{dx}{\sqrt{p^2 - x^2}} = \sin^{-1}\left(\dfrac{x}{p}\right)$ or $-\cos^{-1}\left(\dfrac{x}{p}\right).$

112. $\int \sqrt{ax^2 + c}\, dx = \dfrac{x}{2}\sqrt{ax^2 + c} + \dfrac{c}{2\sqrt{a}}\log\left(x\sqrt{a} + \sqrt{ax^2 + c}\right), a > 0.$

113. $\int \sqrt{ax^2 + c}\, dx = \dfrac{x}{2}\sqrt{ax^2 + c} + \dfrac{c}{2\sqrt{-a}}\sin^{-1}\left(x\sqrt{\dfrac{-a}{c}}\right), a < 0.$

114. $\int \dfrac{dx}{\sqrt{ax^2 + c}} = \dfrac{1}{\sqrt{a}}\log\left(x\sqrt{a} + \sqrt{ax^2 + c}\right), a > 0.$

115. $\int \dfrac{dx}{\sqrt{ax^2 + c}} = \dfrac{1}{\sqrt{-a}}\sin^{-1}\left(x\sqrt{\dfrac{-a}{c}}\right), a < 0.$

116. $\int x\sqrt{ax^2 + c}\, dx = \dfrac{1}{3a}\left(ax^2 + c\right)^{\frac{3}{2}}.$

117. $\int x^2\sqrt{ax^2 + c}\, dx = \dfrac{x}{4a}\sqrt{\left(ax^2 + c\right)^3} - \dfrac{cx}{8a}\sqrt{ax^2 + c}$
$$- \dfrac{c^2}{8\sqrt{a^3}}\log\left(x\sqrt{a} + \sqrt{ax^2 + c}\right), a > 0.$$

118. $\int x^2\sqrt{ax^2 + c}\, dx = \dfrac{x}{4a}\sqrt{\left(ax^2 + c\right)^3} - \dfrac{cx}{8a}\sqrt{ax^2 + c}$
$$- \dfrac{c^2}{8a\sqrt{-a}}\sin^{-1}\left(x\sqrt{\dfrac{-a}{c}}\right), a < 0.$$

119. $\int \dfrac{x\,dx}{\sqrt{ax^2 + c}} = \dfrac{1}{a}\sqrt{ax^2 + c}.$

120. $\int \dfrac{x^2 dx}{\sqrt{ax^2 + c}} = \dfrac{x}{a}\sqrt{ax^2 + c} - \dfrac{1}{a}\int \sqrt{ax^2 + c}\ dx.$

121. $\int \dfrac{\sqrt{ax^2 + c}}{x}dx = \sqrt{ax^2 + c} + \sqrt{c} \log \dfrac{\sqrt{ax^2 + c} - \sqrt{c}}{x}, c > 0.$

122. $\int \sqrt{\dfrac{ax^2 + c}{x}}\ dx = \sqrt{ax^2 + c} - \sqrt{-c} \tan^{-1} \dfrac{\sqrt{ax^2 + c}}{\sqrt{-c}}, c < 0.$

123. $\int \dfrac{dx}{x\sqrt{p^2 \pm x^2}} = -\dfrac{1}{p} \log \left(\dfrac{p + \sqrt{p^2 \pm x^2}}{x} \right).$

124. $\int \dfrac{dx}{x\sqrt{x^2 - p^2}} = \dfrac{1}{p}\cos^{-1}\left(\dfrac{p}{x}\right), \text{ or } -\dfrac{1}{p}\sin^{-1}\left(\dfrac{p}{x}\right).$

125. $\int \dfrac{dx}{x\sqrt{ax^2 + c}} = \dfrac{1}{\sqrt{c}} \log \dfrac{\sqrt{ax^2 + c} - \sqrt{c}}{x}, c > 0.$

126. $\int \dfrac{dx}{x\sqrt{ax^2 + c}} = \dfrac{1}{\sqrt{-c}}\sec^{-1}\left(x\sqrt{-\dfrac{a}{c}}\right), c < 0.$

127. $\int \dfrac{dx}{x^2\sqrt{ax^2 + c}} = -\dfrac{\sqrt{ax^2 + c}}{cx}.$

128. $\int \dfrac{x^n dx}{\sqrt{ax^2 + c}} = \dfrac{x^{n-1}\sqrt{ax^2 + c}}{na} - \dfrac{(n-1)c}{na}\int \dfrac{x^{n-2}dx}{\sqrt{ax^2 + c}}, n > 0.$

129. $\int x^n \sqrt{ax^2 + c}\ dx = \dfrac{x^{n-1}\left(ax^2 + c\right)^{\frac{3}{2}}}{(n+2)a} - \dfrac{(n-1)c}{(n+2)a}$
$$\int x^{n-2}\sqrt{ax^2 + c}\ dx, n > 0.$$

130. $\int \dfrac{\sqrt{ax^2 + c}}{x^n}\ dx = -\dfrac{\left(ax^2 + c\right)^{\frac{3}{2}}}{c(n-1)x^{n-2}} - \dfrac{(n-4)a}{(n-1)c}\int \dfrac{\sqrt{ax^2 + c}}{x^{n-2}}\ dx, n > 1.$

131. $\int \dfrac{dx}{x^n \sqrt{ax^2 + c}} = -\dfrac{\sqrt{ax^2 + c}}{c(n-1)x^{n-1}} - \dfrac{(n-2)a}{(n-1)c}\int \dfrac{dx}{x^{n-2}\sqrt{ax^2 + c}}, n > 1.$

132. $\int \left(ax^2 + c\right)^{\frac{3}{2}} dx = \dfrac{x}{8}\left(2ax^2 + 5c\right)\sqrt{ax^2 + c}$

$$+ \dfrac{3c^2}{8\sqrt{a}}\log\left(x\sqrt{a} + \sqrt{ax^2 + c}\right), a > 0.$$

133. $\int \left(ax^2 + c\right)^{\frac{3}{2}} dx = \dfrac{x}{8}\left(2ax^2 + 5c\right)\sqrt{ax^2 + c}$

$$+ \dfrac{3c^2}{8\sqrt{-a}}\sin^{-1}\left(x\sqrt{\dfrac{-a}{c}}\right), a < 0.$$

134. $\int \dfrac{dx}{\left(ax^2 + c\right)^{\frac{3}{2}}} = \dfrac{x}{c\sqrt{ax^2 + c}}.$

135. $\int x\left(ax^2 + c\right)^{\frac{3}{2}} dx = \dfrac{1}{5a}\left(ax^2 + c\right)^{\frac{5}{2}}.$

136. $\int x^2 \left(ax^2 + c\right)^{\frac{3}{2}} dx = \dfrac{x^3}{6}\left(ax^2 + c\right)^{\frac{3}{2}} + \dfrac{c}{2}\int x^2 \sqrt{ax^2 + c}\; dx.$

137. $\int x^n \left(ax^2 + c\right)^{\frac{3}{2}} dx = \dfrac{x^{n+1}\left(ax^2 + c\right)^{\frac{3}{2}}}{n+4} + \dfrac{3c}{n+4}\int x^n \sqrt{ax^2 + c}\; dx.$

138. $\int \dfrac{xdx}{\left(ax^2 + c\right)^{\frac{3}{2}}} = -\dfrac{1}{a\sqrt{ax^2 + c}}.$

139. $\int \dfrac{x^2 dx}{\left(ax^2 + c\right)^{\frac{3}{2}}} = -\dfrac{x}{a\sqrt{ax^2 + c}} + \dfrac{1}{a\sqrt{a}}\log\left(x\sqrt{a} + \sqrt{ax^2 + c}\right), a > 0.$

140. $\displaystyle\int \frac{x^2 dx}{\left(ax^2 + c\right)^{\frac{3}{2}}} = -\frac{x}{a\sqrt{ax^2 + c}} + \frac{1}{a\sqrt{-a}} \sin^{-1}\left(x\sqrt{\frac{-a}{c}}\right), a < 0.$

141. $\displaystyle\int \frac{x^3 dx}{\left(ax^2 + c\right)^{\frac{3}{2}}} = -\frac{x^2}{a\sqrt{ax^2 + c}} + \frac{2}{a^2}\sqrt{ax^2 + c}.$

142. $\displaystyle\int \frac{dx}{x\left(ax^n + c\right)} = \frac{1}{cn} \log \frac{x^n}{ax^n + c}.$

143. $\displaystyle\int \frac{dx}{\left(ax^n + c\right)^m} = \frac{1}{c}\int \frac{dx}{\left(ax^n + c\right)^{m-1}} - \frac{a}{c}\int \frac{x^n dx}{\left(ax^n + c\right)^m}.$

144. $\displaystyle\int \frac{dx}{x\sqrt{ax^n + c}} = \frac{1}{n\sqrt{c}} \log \frac{\sqrt{ax^n + c} - \sqrt{c}}{\sqrt{ax^n + c} + \sqrt{c}}, c > 0.$

145. $\displaystyle\int \frac{dx}{x\sqrt{ax^n + c}} = \frac{2}{n\sqrt{-c}} \sec^{-1}\sqrt{\frac{-ax^n}{c}}, c < 0.$

146. $\displaystyle\int x^{m-1}\left(ax^n + c\right)^p dx$

$\displaystyle = \frac{1}{m + np}\left[x^m\left(ax^n + c\right)^p + npc\int x^{m-1}\left(ax^n + c\right)^{p-1} dx\right].$

$\displaystyle = \frac{1}{cn(p + 1)}\left[-x^m\left(ax^n + c\right)^{p+1} + (m + np + n)\int x^{m-1}\left(ax^n + c\right)^{p+1} dx\right].$

$\displaystyle = \frac{1}{a(m + np)}\left[x^{m-n}\left(ax^n + c\right)^{p+1} - (m - n)c\int x^{m-n-1}\left(ax^n + c\right)^p dx\right].$

$\displaystyle = \frac{1}{mc}\left[x^m\left(ax^n + c\right)^{p+1} - (m + np + n)a\int x^{m+n-1}\left(ax^n + c\right)^p dx\right].$

147. $\displaystyle\int \frac{x^m dx}{\left(ax^n + c\right)^p} = \frac{1}{a}\int \frac{x^{m-n} dx}{\left(ax^n + c\right)^{p-1}} - \frac{c}{a}\int \frac{x^{m-n} dx}{\left(ax^n + c\right)^p}.$

148. $\displaystyle\int \frac{dx}{x^m\left(ax^n + c\right)^p} = \frac{1}{c}\int \frac{dx}{x^m\left(ax^n + c\right)^{p-1}} - \frac{a}{c}\int \frac{dx}{x^{m-n}\left(ax^n + c\right)^p}.$

Expressions Containing $(ax^2 + bx + c)$.

149. $\displaystyle \int \frac{dx}{ax^2 + bx + c} = \frac{1}{\sqrt{b^2 - 4ac}} \log \frac{2ax + b - \sqrt{b^2 - 4ac}}{2ax + b + \sqrt{b^2 - 4ac}}, b^2 > 4ac.$

150. $\displaystyle \int \frac{dx}{ax^2 + bx + c} = \frac{2}{\sqrt{4ac - b^2}} \tan^{-1} \frac{2ax + b}{\sqrt{4ac - b^2}}, b^2 < 4ac.$

151. $\displaystyle \int \frac{dx}{ax^2 + bx + c} = -\frac{2}{2ax + b}, b^2 = 4ac.$

152. $\displaystyle \int \frac{dx}{\left(ax^2 + bx + c\right)^{n+1}} = \frac{2ax + b}{n\left(4ac - b^2\right)\left(ax^2 + bx + c\right)^n}$
$$+ \frac{2(2n - 1)a}{n\left(4ac - b^2\right)} \int \frac{dx}{\left(ax^2 + bx + c\right)^n}.$$

153. $\displaystyle \int \frac{xdd}{ax^2 + bx + c} = \frac{1}{2a} \log\left(ax^2 + bx + c\right) - \frac{b}{2a} \int \frac{dx}{ax^2 + bx + c}.$

154. $\displaystyle \int \frac{x^2 dx}{ax^2 + bx + c} = \frac{x}{a} - \frac{b}{2a^2} \log\left(ax^2 + bx + c\right)$
$$+ \frac{b^2 - 2ac}{2a^2} \int \frac{dx}{ax^2 + bx + c}.$$

155. $\displaystyle \int \frac{x^n dx}{ax^2 + bx + c} = \frac{x^{n-1}}{(n-1)a} - \frac{c}{a} \int \frac{x^{n-2} dx}{ax^2 + bx + c} - \frac{b}{a} \int \frac{x^{n-1} dx}{ax^2 + bx + c}.$

156. $\displaystyle \int \frac{xdx}{\left(ax^2 + bx + c\right)^{n+1}} = \frac{-(2c + bx)}{n\left(4ac - b^2\right)\left(ax^2 + bx + c\right)^n}$
$$- \frac{b(2n - 1)}{n\left(4ac - b^2\right)} \int \frac{dx}{\left(ax^2 + bx + c\right)^n}.$$

157. $\displaystyle\int \frac{x^m\,dx}{\left(ax^2+bx+c\right)^{n+1}} = -\frac{x^{m-1}}{a(2n-m+1)\left(ax^2+bx+c\right)^n}$

$$-\frac{(n-m+1)}{(2n-m+1)}\cdot\frac{b}{a}\int\frac{x^{m-1}\,dx}{\left(ax^2+bx+c\right)^{n+1}}$$

$$+\frac{(m-1)}{(2n-m+1)}\cdot\frac{c}{a}\int\frac{x^{m-2}}{\left(ax^2+bx+c\right)^{n+1}}.$$

158. $\displaystyle\int \frac{dx}{x\left(ax^2+bx+c\right)} = \frac{1}{2c}\log\left(\frac{x^2}{\left(ax^2+bx+c\right)}\right) - \frac{b}{2c}\int\frac{dx}{\left(ax^2+bx+c\right)}.$

159. $\displaystyle\int \frac{dx}{x^2\left(ax^2+bx+c\right)} = \frac{b}{2c^2}\log\left(\frac{ax^2+bx+c}{x^2}\right) - \frac{1}{cx}$

$$+\left(\frac{b^2}{2c^2}-\frac{a}{c}\right)\int\frac{dx}{\left(ax^2+bx+c\right)}.$$

160. $\displaystyle\int \frac{dx}{x^m\left(ax^2+bx+c\right)^{n+1}} = -\frac{1}{(m-1)cx^{m-1}\left(ax^2+bx+c\right)^n}$

$$-\frac{(n+m-1)}{m-1}\cdot\frac{b}{c}\int\frac{dx}{x^{m-1}\left(ax^2+bx+c\right)^{n+1}}$$

$$-\frac{(2n+m-1)}{m-1}\cdot\frac{a}{c}\int\frac{dx}{x^{m-2}\left(ax^2+bx+c\right)^{n+1}}.$$

161. $\displaystyle\int \frac{dx}{x\left(ax^2+bx+c\right)^n} = \frac{1}{2c(n-1)\left(ax^2+bx+c\right)^{n-1}}$

$$-\frac{b}{2c}\int\frac{dx}{\left(ax^2+bx+c\right)^n}+\frac{1}{c}\int\frac{dx}{x\left(ax^2+bx+c\right)^{n-1}}.$$

162. $\displaystyle\int \frac{dx}{\sqrt{ax^2+bx+c}} = \frac{1}{\sqrt{a}}\log\left(2ax+b+2\sqrt{a}\sqrt{ax^2+bx+c}\right),\ a>0.$

163. $\int \dfrac{dx}{\sqrt{ax^2 + bx + c}} = \dfrac{1}{\sqrt{-a}} \sin^{-1} \dfrac{-2ax - b}{\sqrt{b^2 - 4ac}}, a < 0.$

164. $\int \dfrac{x\,dx}{\sqrt{ax^2 + bx + c}} = \dfrac{\sqrt{ax^2 + bx + c}}{a} - \dfrac{b}{2a} \dfrac{dx}{\sqrt{ax^2 + bx + c}}.$

165. $\int \dfrac{x^n\,dx}{\sqrt{ax^2 + bx + c}} = \dfrac{x^{n-1}}{am} \sqrt{ax^2 + bx + c} - \dfrac{b(2n - 1)}{2an}$

$$\int \dfrac{x^{n-1}\,dx}{\sqrt{ax^2 + bx + c}} - \dfrac{c(n - 1)}{an} \int \dfrac{x^{n-2}\,dx}{\sqrt{ax^2 + bx + c}}.$$

166. $\int \sqrt{ax^2 + bx + c}\, dx = \dfrac{2ax + b}{4a} \sqrt{ax^2 + bx + c}$

$$+ \dfrac{4ac - b^2}{8a} \int \dfrac{dx}{\sqrt{ax^2 + bx + c}}.$$

167. $\int x\sqrt{ax^2 + bx + c}\, dx = \dfrac{\left(ax^2 + bx + c\right)^{\frac{3}{2}}}{3a} - \dfrac{b}{2a} \int \sqrt{ax^2 + bx + c}\, dx.$

168. $\int x^2 \sqrt{ax^2 + bx + c}\, dx = \left(x - \dfrac{5b}{6a}\right) \dfrac{\left(ax^2 + bx + c\right)^{\frac{3}{2}}}{4a}$

$$+ \dfrac{\left(5b^2 - 4ac\right)}{16a^2} \int \sqrt{ax^2 + bx + c}\, dx.$$

169. $\int \dfrac{dx}{x\sqrt{ax^2 + bx + c}} = -\dfrac{1}{\sqrt{c}} \log\left(\dfrac{\sqrt{ax^2 + bx + c} + \sqrt{c}}{x} + \dfrac{b}{2\sqrt{c}}\right), c > 0.$

170. $\int \dfrac{dx}{x\sqrt{ax^2 + bx + c}} = \dfrac{1}{\sqrt{-c}} \sin^{-1} \dfrac{bx + 2c}{x\sqrt{b^2 - 4ac}}, c < 0.$

171. $\int \dfrac{dx}{x\sqrt{ax^2 + bx}} = -\dfrac{2}{bx} \sqrt{ax^2 + bx}, c = 0.$

172. $\int \dfrac{dx}{x^n \sqrt{ax^2 + bx + c}} = -\dfrac{\sqrt{ax^2 + bx + c}}{c(n-1)x^{n-1}}$

$+ \dfrac{b(3-2n)}{2c(n-1)} \int \dfrac{dx}{x^{n-1}\sqrt{ax^2+bx+c}} + \dfrac{a(2-n)}{c(n-1)} \int \dfrac{dx}{x^{n-2}\sqrt{ax^2+bx+c}}.$

173. $\int \dfrac{dx}{\left(ax^2 + bx + c\right)^{\frac{3}{2}}} = -\dfrac{2(2ax+b)}{\left(b^2 - 4ac\right)\sqrt{ax^2+bx+c}}, b^2 \neq 4ac.$

174. $\int \dfrac{dx}{\left(ax^2 + bx + c\right)^{\frac{3}{2}}} = -\dfrac{1}{2\sqrt{a^3}\left(x+b/2a\right)^2}, b^2 = 4ac.$

Miscellaneous Algebraic Expressions

175. $\int \sqrt{2px - x^2}\; dx = \dfrac{1}{2}\left[(x-p)\sqrt{2px - x^2} + p^2 \sin^{-1}\left[(x-p)/p\right]\right].$

176. $\int \dfrac{dx}{\sqrt{2px - x^2}} = \cos^{-1}\left(\dfrac{p-x}{p}\right).$

177. $\int \dfrac{dx}{\sqrt{ax + b}.\sqrt{cx + d}} = \dfrac{2}{\sqrt{-ac}} \tan^{-1}\sqrt{\dfrac{-c(ax+b)}{a(cx+d)}},$

$\text{or } \dfrac{2}{\sqrt{ac}} \tanh^{-1}\sqrt{\dfrac{c(ax+b)}{a(cx+d)}}.$

178. $\int \sqrt{ax + b}.\sqrt{cx + d}\; dx = \dfrac{(2acx + bc + ad)\sqrt{ax+b}.\sqrt{cx+d}}{4ac}$

$-\dfrac{(ad-bc)^2}{8ac} \int \dfrac{dx}{\sqrt{ax+b}.\sqrt{cx+d}}.$

179. $\int \sqrt{\dfrac{cx+d}{ax+b}}\; dx = \dfrac{\sqrt{ax+b}.\sqrt{cx+d}}{a} + \dfrac{(ad-bc)}{2a} \int \dfrac{dx}{\sqrt{ax+b}.\sqrt{cx+d}}.$

180. $\int \sqrt{\dfrac{x+b}{x+d}}\; dx = \sqrt{x+d}.\sqrt{x+b} + (b-d)\log\left[\sqrt{x+d} + \sqrt{x+b}\right].$

181. $\int \sqrt{\dfrac{1+x}{1-x}}\, dx = \sin^{-1} x - \sqrt{1-x^2}$.

182. $\int \sqrt{\dfrac{p-x}{q+x}}\, dx = \sqrt{p-x}.\sqrt{q+x} + (p+q)\sin^{-1}\sqrt{\dfrac{x+q}{p+q}}$.

183. $\int \sqrt{\dfrac{p+x}{q-x}}\, dx = -\sqrt{p+x}.\sqrt{q-x} - (p+q)\sin^{-1}\sqrt{\dfrac{q-x}{p+q}}$.

184. $\int \dfrac{dx}{\sqrt{x-p}.\sqrt{q-x}} = 2\sin^{-1}\sqrt{\dfrac{x-p}{q-p}}$.

Expressions Containing sin *ax*

185. $\int \sin u\, du = -\cos u$, where u is any function of x.

186. $\int \sin ax\, dx = -\dfrac{1}{a}\cos ax$.

187. $\int \sin^2 ax\, dx = \dfrac{x}{2} - \dfrac{\sin 2ax}{4a}$.

188. $\int \sin^3 ax\, dx = -\dfrac{1}{a}\cos ax + \dfrac{1}{3a}\cos^3 ax$.

189. $\int \sin^4 ax\, dx = \dfrac{3x}{8} - \dfrac{3\sin 2ax}{16a} - \dfrac{\sin^3 ax \cos ax}{4a}$.

190. $\int \sin^n ax\, dx = -\dfrac{\sin^{n-1} ax \cos ax}{na} + \dfrac{n-1}{n}\int \sin^{n-2} ax\, dx$,
(n pos. integer).

191. $\int \dfrac{dx}{\sin ax} = \dfrac{1}{a}\log\tan\dfrac{ax}{2} = \dfrac{1}{a}\log(\csc ax - \operatorname{ctn} ax)$.

192. $\int \dfrac{dx}{\sin^2 ax} = \int \csc^2 ax \, dx = -\dfrac{1}{a} \operatorname{ctn} ax.$

193. $\int \dfrac{dx}{\sin^n ax} = -\dfrac{1}{a(n-1)} \dfrac{\cos ax}{\sin^{n-1} ax} + \dfrac{n-2}{n-1} \int \dfrac{dx}{\sin^{n-2} ax},$

n integer > 1.

194. $\int \dfrac{dx}{1 \pm \sin ax} = \mp \dfrac{1}{a} \tan \left(\dfrac{\pi}{4} \mp \dfrac{ax}{2} \right).$

195. $\int \dfrac{dx}{b + c \sin ax} = \dfrac{-2}{a\sqrt{b^2 - c^2}} \tan^{-1} \left[\sqrt{\dfrac{b-c}{b+c}} \tan \left(\dfrac{\pi}{4} - \dfrac{ax}{2} \right) \right], b^2 > c^2.$

196. $\int \dfrac{dx}{b + c \sin ax} = \dfrac{-1}{a\sqrt{c^2 - b^2}} \log \dfrac{c + b \sin ax + \sqrt{c^2 - b^2} \, \cos ax}{b + c \sin ax}, c^2 > b^2.$

197. $\int \sin ax \sin bx \, dx = \dfrac{\sin(a-b)x}{2(a-b)} - \dfrac{\sin(a+b)x}{2(a+b)}, a^2 \neq b^2.$

198. $\int \sqrt{1 + \sin x} \, dx = \pm 2 \left(\sin \dfrac{x}{2} - \cos \dfrac{x}{2} \right);$ use + sign

when $(8k-1)\dfrac{\pi}{2} < x \leq (8k+3)\dfrac{\pi}{2}$, otherwise –, k an integer.

199. $\int \sqrt{1 - \sin x} \, dx = \pm 2 \left(\sin \dfrac{x}{2} + \cos \dfrac{x}{2} \right);$ use + sign

when $(8k-3)\dfrac{\pi}{2} < x \leq (8k+1)\dfrac{\pi}{2}$, otherwise –, k an integer.

Expressions Involving cos ax

200. $\int \cos u \, du = \sin u,$ where u is any function of x.

201. $\int \cos ax \, dx = \dfrac{1}{a} \sin ax.$

202. $\int \cos^2 ax \, dx = \dfrac{x}{2} + \dfrac{\sin 2ax}{4a}.$

203. $\int \cos^3 ax \, dx = \dfrac{1}{a} \sin ax - \dfrac{1}{3} \sin^3 ax.$

204. $\int \cos^4 ax \, dx = \dfrac{3x}{8} + \dfrac{3 \sin 2ax}{16a} + \dfrac{\cos^3 ax \sin ax}{4a}.$

205. $\int \cos^n ax \, dx = \dfrac{\cos^{n-1} ax \sin ax}{na} + \dfrac{n-1}{n} \int \cos^{n-2} ax \, dx.$

206. $\int \dfrac{dx}{\cos ax} = \dfrac{1}{a} \log \tan \left(\dfrac{ax}{2} + \dfrac{\pi}{4} \right) = \dfrac{1}{a} \log(\tan ax + \sec ax).$

207. $\int \dfrac{dx}{\cos^2 ax} = \dfrac{1}{a} \tan ax.$

208. $\int \dfrac{dx}{\cos^n ax} = \dfrac{1}{a(n-1)} \dfrac{\sin ax}{\cos^{n-1} ax} + \dfrac{n-2}{n-1} \int \dfrac{dx}{\cos^{n-2} ax},$
n integer > 1.

209. $\int \dfrac{dx}{1 + \cos ax} = \dfrac{1}{a} \tan \dfrac{ax}{2}.$

210. $\int \dfrac{dx}{1 - \cos ax} = -\dfrac{1}{a} \cot \dfrac{ax}{2}.$

211. $\int \sqrt{1 + \cos x} \, dx = \pm\sqrt{2} \int \cos \dfrac{x}{2} \, dx = \pm 2\sqrt{2} \sin \dfrac{x}{2}.$
Use + when $(4k - 1) \pi < x \leq (4k + 1) \pi$, otherwise −, k an integer.

212. $\int \sqrt{1 - \cos x} \; dx = \pm\sqrt{2} \int \sin \dfrac{x}{2} \, dx = \mp 2\sqrt{2} \, \cos \dfrac{x}{2}.$

Use top signs when $4k\pi < x \leqq (4k + 2)\pi$, otherwise bottom signs.

213. $\int \dfrac{dx}{b + c \cos ax} = \dfrac{1}{a\sqrt{b^2 - c^2}} \tan^{-1}\left(\dfrac{\sqrt{b^2 - c^2} . \sin ax}{c + b \cos ax} \right), b^2 > c^2.$

214. $\int \dfrac{dx}{b + c \cos ax} = \dfrac{1}{a\sqrt{c^2 - b^2}} \tanh^{-1}\left[\dfrac{\sqrt{c^2 - b^2} . \sin ax}{c + b \cos ax} \right], c^2 > b^2.$

215. $\int \cos ax \cdot \cos bx \; dx = \dfrac{\sin(a - b)x}{2(a - b)} + \dfrac{\sin(a + b)x}{2(a + b)}, a^2 \neq b^2.$

Expressions Containing sin ax and cos ax

216. $\int \sin ax \cos bx \; dx = -\dfrac{1}{2}\left[\dfrac{\cos(a - b)x}{a - b} + \dfrac{\cos(a + b)x}{a + b} \right], a^2 \neq b^2.$

217. $\int \sin^n ax \cos ax \; dx = \dfrac{1}{a(n + 1)} \sin^{n+1} ax, n \neq -1.$

218. $\int \cos^n ax \sin ax \; dx = -\dfrac{1}{a(n + 1)} \cos^{n+1} ax, n \neq -1.$

219. $\int \dfrac{\sin ax}{\cos ax} \, dx = -\dfrac{1}{a} \log \cos ax.$

220. $\int \dfrac{\cos ax}{\sin ax} \, dx = \dfrac{1}{a} \log \sin ax.$

221. $\int (b + c \sin ax)^n \cos ax \; dx = \dfrac{1}{ac(n + 1)} (b + c \sin ax)^{n+1}, n \neq -1.$

222. $\int (b + c \cos ax)^n \sin ax \; dx = -\dfrac{1}{ac(n + 1)} (b + c \cos ax)^{n+1}, n \neq -1.$

223. $\int \dfrac{\cos ax \, dx}{b + c \sin ax} = \dfrac{1}{ac} \log(b + c \sin ax).$

224. $\int \dfrac{\sin ax}{b + c \cos ax} \, dx = -\dfrac{1}{ac} \log(b + c \cos ax).$

225. $\int \dfrac{dx}{b \sin ax + c \cos ax} = \dfrac{1}{a\sqrt{b^2 + c^2}} \left[\log \tan \dfrac{1}{2} \left(ax + \tan^{-1} \dfrac{c}{b} \right) \right].$

226. $\int \dfrac{dx}{b + c \cos ax + d \sin ax} = \dfrac{-1}{a\sqrt{b^2 - c^2 - d^2}} \sin^{-1} U,$

$$U \equiv \left[\dfrac{c^2 + d^2 + b(c \cos ax + d \sin ax)}{\sqrt{c^2 + d^2} \, (b + c \cos ax + d \sin ax)} \right]; \text{ or } = \dfrac{1}{a\sqrt{c^2 + d^2 - b^2}} \log V,$$

$$V \equiv \left[\dfrac{c^2 + d^2 + b(c \cos ax + d \sin ax) + \sqrt{c^2 + d^2 - b^2} \, (c \sin ax - d \cos ax)}{\sqrt{c^2 + d^2} \, (b + c \cos ax + d \sin ax)} \right]$$

$$b^2 \neq c^2 + d^2, -\pi < ax < \pi.$$

227. $\int \dfrac{dx}{b + c \cos ax + d \sin ax}$

$$= \dfrac{1}{ab} \left[\dfrac{b - (c + d)\cos ax + (c - d)\sin ax}{b + (c - d)\cos ax + (c + d)\sin ax} \right], b^2 = c^2 + d^2.$$

228. $\int \dfrac{\sin^2 ax \, dx}{b + c \cos^2 ax} = \dfrac{1}{ac} \sqrt{\dfrac{b + c}{b}} \tan^{-1} \left(\sqrt{\dfrac{b}{b + c}} \cdot \tan ax \right) - \dfrac{x}{c}.$

229. $\int \dfrac{\sin ax \cos ax \, dx}{b \cos^2 ax + c \sin^2 ax} = \dfrac{1}{2a(c - b)} \log\left(b \cos^2 ax + c \sin^2 ax\right).$

230. $\int \dfrac{dx}{b^2 \cos^2 ax - c^2 \sin^2 ax} = \dfrac{1}{2abc} \log \dfrac{b \cos ax + c \sin ax}{b \cos ax - c \sin ax}.$

231. $\displaystyle\int \frac{dx}{b^2 \cos^2 ax + c^2 \sin^2 ax} = \frac{1}{abc} \tan^{-1}\left(\frac{c \tan ax}{b}\right).$

232. $\displaystyle\int \sin^2 ax \cos^2 ax \, dx = \frac{x}{8} - \frac{\sin 4ax}{32a}.$

233. $\displaystyle\int \frac{dx}{\sin ax \cos ax} = \frac{1}{a} \log \tan ax.$

234. $\displaystyle\int \frac{dx}{\sin^2 ax \cos^2 ax} = \frac{1}{a}(\tan ax - \cot ax).$

235. $\displaystyle\int \frac{\sin^2 ax}{\cos ax} \, dx = \frac{1}{a}\left[-\sin ax + \log \tan\left(\frac{ax}{2} + \frac{\pi}{4}\right)\right].$

236. $\displaystyle\int \frac{\cos^2 ax}{\sin ax} \, dx = \frac{1}{a}\left[\cos ax + \log \tan \frac{ax}{2}\right].$

237. $\displaystyle\int \sin^m ax \cos^n ax \, dx = -\frac{\sin^{m-1} ax \cos^{n+1} ax}{a(m+n)}$
$$+ \frac{m-1}{m+n}\int \sin^{m-2} ax \cos^n ax \, dx, m, n > 0.$$

238. $\displaystyle\int \sin^m ax \cos^n ax \, dx = \frac{\sin^{m+1} ax \cos^{n-1} ax}{a(m+n)} + \frac{n-1}{m+n}$
$$\int \sin^m ax \cos^{n-2} ax \, dx, m, n > 0.$$

239. $\displaystyle\int \frac{\sin^m ax}{\cos^n ax} \, dx = \frac{\sin^{m+1} ax}{a(n-1)\cos^{n-1} ax} - \frac{m-n+2}{n-1}$
$$\int \frac{\sin^m ax}{\cos^{n-2} ax} \, dx, m, n > 0, n \neq 1.$$

240. $\displaystyle\int \frac{\cos^n ax}{\sin^m ax} \, dx = \frac{-\cos^{n+1} ax}{a(m-1)\sin^{m-1} ax} + \frac{m-n-2}{(m-1)}$
$$\int \frac{\cos^n ax}{\sin^{m-2} ax} \, dx, m, n, > 0, m \neq 1.$$

241. $\displaystyle\int\frac{dx}{\sin^m ax \cos^n ax} = \frac{1}{a(n-1)}\frac{1}{\sin^{m-1}\cos^{n-1} ax} + \frac{m+n-2}{(n-1)}$

$$\int\frac{dx}{\sin^m ax \cos^{n-2} ax}.$$

242. $\displaystyle\int\frac{dx}{\sin^m ax \cos^n ax} = -\frac{1}{a(m-1)}\frac{1}{\sin^{m-1} ax \cos^{n-1} ax}$

$$+\frac{m+n-2}{(m-1)}\int\frac{dx}{\sin^{m-2} ax \cos^n ax}.$$

243. $\displaystyle\int\frac{\sin^{2n} ax}{\cos ax}dx = \int\frac{\left(1-\cos^2 ax\right)^n}{\cos ax}dx.$ (Expand, divide, and use Art. 205).

244. $\displaystyle\int\frac{\cos^{2n} ax}{\sin ax}dx = \int\frac{\left(1-\sin^2 ax\right)^n}{\sin ax}dx.$ (Expand, divide, and use Art. 190).

245. $\displaystyle\int\frac{\sin^{2n+1} ax}{\cos ax}dx = \int\frac{\left(1-\cos^2 ax\right)^n}{\cos ax}\sin ax\, dx.$

(Expand, divide, and use Art. 218).

246. $\displaystyle\int\frac{\cos^{2n+1} ax}{\sin ax}dx = \int\frac{\left(1-\sin^2 ax\right)^n}{\sin ax}\cos ax\, dx.$

(Expand, divide, and use Art. 217).

Expressions Containing **tan *ax* or cot *ax* (tan *ax*=1/cot *ax*)**

247. $\displaystyle\int\tan u\, du = -\log\cos u,$ or $\log\sec u,$ where u is any function

of x.

248. $\displaystyle\int\tan ax\, dx = -\frac{1}{a}\log\cos ax.$

249. $\int \tan^2 ax \, dx = \dfrac{1}{a} \tan ax - x.$

250. $\int \tan^3 ax \, dx = \dfrac{1}{2a} \tan^2 ax + \dfrac{1}{a} \log \cos ax.$

251. $\int \tan^n ax \, dx = \dfrac{1}{a(n-1)} \tan^{n-1} ax - \int \tan^{n-2} ax \, dx, \; n \text{ integer} > 1.$

252. $\int \cot u \, du = \log \sin u, \;$ or $\; -\log \; \mathrm{cosec} \; u, \;$ where $\; u \;$ is any function of $x.$

253. $\int \cot^2 ax \, dx = \int \dfrac{dx}{\tan^2 ax} = -\dfrac{1}{a} \cot ax - x.$

254. $\int \cot^3 ax \, dx = -\dfrac{1}{2a} \cot^2 ax - \dfrac{1}{a} \log \sin ax.$

255. $\int \cot^n ax \, dx = \int \dfrac{dx}{\tan^n ax} = -\dfrac{1}{a(n-1)} ax - \int \cot^{n-2} ax \, dx,$

$n \text{ integer} > 1.$

256. $\int \dfrac{dx}{b + c \tan ax} = \int \dfrac{\cot ax \, dx}{b \cot ax + c} = \dfrac{1}{b^2 + c^2}$

$= \dfrac{1}{b^2 + c^2} \left[bx + \dfrac{c}{a} \log(b \cos ax + c \sin ax) \right]$

257. $\int \dfrac{dx}{b + c \cot ax} = \int \dfrac{\tan ax \, dx}{b \tan ax + c}$

$= \dfrac{1}{b^2 + c^2} \left[bx - \dfrac{c}{a} \log(c \cos ax + b \sin ax) \right].$

258. $\int \dfrac{dx}{\sqrt{b + c \tan^2 ax}} = \dfrac{1}{a\sqrt{b-c}} \sin^{-1} \left(\sqrt{\dfrac{b-c}{b}} \sin ax \right),$

$b \text{ pos.}, \; b^2 > c^2.$

*Expressions Containing **sec** ax = 1/cos ax or cosec ax = 1 sin ax.*

259. $\int \sec u \, du = \log(\sec u + \tan u) = \log \tan \left(\dfrac{u}{2} + \dfrac{\pi}{4} \right)$, where u is

any function of x.

260. $\int \sec ax \, dx = \dfrac{1}{a} \log \tan \left(\dfrac{ax}{2} + \dfrac{\pi}{4} \right)$.

261. $\int \sec^2 ax \, dx = \dfrac{1}{a} \tan ax$.

262. $\int \sec^3 ax \, dx = \dfrac{1}{2a} \left[\tan ax \sec ax + \log \tan \left(\dfrac{ax}{2} + \dfrac{\pi}{4} \right) \right]$.

263. $\int \sec^n ax \, dx = \dfrac{1}{a(n-1)} \dfrac{\sin ax}{\cos^{n-1} ax} + \dfrac{n-2}{n-1} \int \sec^{n-2} ax \, dx$,

n integer > 1.

264. $\int \operatorname{cosec} u \, du = \log(\operatorname{cosec} u - \cot u) = \log \tan \dfrac{u}{2}$, where u is

any function of x.

265. $\int \operatorname{cosec} ax \, dx = \dfrac{1}{a} \log \tan \dfrac{ax}{2}$.

266. $\int \operatorname{cosec}^2 ax \, dx = -\dfrac{1}{a} \cot ax$.

267. $\int \operatorname{cosec}^3 ax \, dx = \dfrac{1}{2a} \left[-\cot ax \operatorname{cosec} ax + \log \tan \dfrac{ax}{2} \right]$.

268. $\int \operatorname{cosec}^n ax \, dx = -\dfrac{1}{a(n-1)} \dfrac{\cos ax}{\sin^{n-1} ax} + \dfrac{n-2}{n-1} \int \operatorname{cosec}^{n-2} ax \, dx$,

n integer > 1.

Expressions Containing tan *ax* and sec *ax* or cot *ax* and cosec *ax*.

269. $\int \tan u \sec u \, du = \sec u$, where u is any function of x.

270. $\int \tan ax \sec ax \, dx = \dfrac{1}{a} \sec ax$.

271. $\int \tan^n ax \sec^2 ax \, dx = \dfrac{1}{a(n+1)} \tan^{n+1} ax, \; n \neq -1$.

272. $\int \tan ax \sec^n ax \, dx = \dfrac{1}{an} \sec^n ax, \; n \neq 0$.

273. $\int \cot u \operatorname{cosec} u \, du = -\operatorname{cosec} u$, where u is any function of x.

274. $\int \cot ax \operatorname{cosec} ax \, dx = -\dfrac{1}{a} \operatorname{cosec} ax$.

275. $\int \cot^n ax \operatorname{co\,sec}^2 ax \, dx = -\dfrac{1}{a(n+1)} \cot^{n+1} ax, \; n \neq -1$.

276. $\int \cot ax \operatorname{cosec}^n ax \, dx = -\dfrac{1}{an} \operatorname{cosec}^n ax, \; n \neq 0$.

277. $\int \dfrac{\operatorname{cosec}^2 ax \, dx}{\cot ax} = -\dfrac{1}{a} \log \cot ax$.

Expressions Containing Algebraic and Trigonometric Functions

278. $\int x \sin ax \, dx = \dfrac{1}{a^2} \sin ax - \dfrac{1}{a} x \cos ax$.

279. $\int x^2 \sin ax \, dx = \dfrac{2x}{a^2} \sin ax + \dfrac{2}{a^3} \cos ax - \dfrac{x^2}{a} \cos ax$.

280. $\int x^3 \sin ax \, dx = \dfrac{3x^2}{a^2} \sin ax - \dfrac{6}{a^4} \sin ax - \dfrac{x^3}{a} \cos ax + \dfrac{6x}{a^3} \cos ax$.

281. $\int x \sin^2 ax \, dx = \dfrac{x^2}{4} - \dfrac{x \sin 2ax}{4a} - \dfrac{\cos 2ax}{8a^2}$

282. $\int x^2 \sin^2 ax \, dx = \dfrac{x^3}{6} - \left(\dfrac{x^2}{4a} - \dfrac{1}{8a^3} \right) \sin 2ax - \dfrac{x \cos 2ax}{4a^2}$

283. $\int x^3 \sin^2 ax \, dx = \dfrac{x^4}{8} - \left(\dfrac{x^3}{4a} - \dfrac{3x}{8a^3} \right) \sin 2ax - \left(\dfrac{3x^2}{8a^2} - \dfrac{3}{16a^4} \right) \cos 2ax.$

284. $\int x \sin^3 ax \, dx = \dfrac{x \cos 3ax}{12a} - \dfrac{\sin 3ax}{36a^2} - \dfrac{3x \cos ax}{4a} + \dfrac{3 \sin ax}{4a^2}.$

285. $\int x^n \sin ax \, dx = -\dfrac{1}{a} x^n \cos ax + \dfrac{n}{a} \int x^{n-1} \cos ax \, dx.$

286. $\int \dfrac{\sin ax \, dx}{x} = ax - \dfrac{(ax)^3}{3 \cdot 3!} + \dfrac{(ax)^5}{5 \cdot 5!} - \ldots.$

287. $\int \dfrac{\sin ax \, dx}{x^m} = \dfrac{-1}{(m-1)} \dfrac{\sin ax}{x^{m-1}} + \dfrac{a}{(m-1)} \int \dfrac{\cos ax \, dx}{x^{m-1}}.$

288. $\int x \cos ax \, dx = \dfrac{1}{a^2} \cos ax + \dfrac{1}{a} x \sin ax.$

289. $\int x^2 \cos ax \, dx = \dfrac{2x}{a^2} \cos ax - \dfrac{2}{a^3} \sin ax + \dfrac{x^2}{a} \sin ax.$

290. $\int x^3 \cos ax \, dx = \dfrac{\left(3a^2 x^2 - 6 \right) \cos ax}{a^4} + \dfrac{\left(a^2 x^3 - 6x \right) \sin ax}{a^3}.$

291. $\int x \cos^2 ax \, dx = \dfrac{x^2}{4} + \dfrac{x \sin 2ax}{4a} + \dfrac{\cos 2ax}{8a^2}.$

292. $\int x^2 \cos^2 ax \, dx = \dfrac{x^3}{6} + \left(\dfrac{x^2}{4a} - \dfrac{1}{8a^3} \right) \sin 2ax + \dfrac{x \cos 2ax}{4a^2}$.

293. $\int x^3 \cos^2 ax \, dx = \dfrac{x^4}{8} + \left(\dfrac{x^3}{4a} - \dfrac{3x}{8a^3} \right) \sin 2ax + \left(\dfrac{3x^2}{8a^2} - \dfrac{3}{16a^4} \right) \cos 2ax.$

294. $\int x \cos^3 ax \, dx = \dfrac{x \sin 3ax}{12a} + \dfrac{\cos 3ax}{36a^2} + \dfrac{3x \sin ax}{4a} + \dfrac{3 \cos ax}{4a^2}$.

295. $\int x^n \cos ax \, dx = \dfrac{1}{a} x^n \sin ax - \dfrac{n}{a} \int x^{n-1} \sin ax \, dx,$ n pos.

296. $\int \dfrac{\cos ax \, dx}{x} = \log ax - \dfrac{(ax)^2}{2 \cdot 2!} + \dfrac{(ax)^4}{4 \cdot 4!} - \ldots$

297. $\int \dfrac{\cos ax}{x^m} \, dx = -\dfrac{1}{(m-1)} \cdot \dfrac{\cos ax}{x^{m-1}} - \dfrac{a}{(m-1)} \int \dfrac{\sin ax \, dx}{x^{m-1}}$.

Expressions Containing Exponential and Logarithmic Functions

298. $\int e^u \, du = e^u$, where u is any function of x.

299. $\int b^u \, du = \dfrac{b^u}{\log b}$, where u is any function of x.

300. $\int e^{ax} \, dx = \dfrac{1}{a} e^{ax}$, $\quad \int b^{ax} \, dx = \dfrac{b^{ax}}{a \log b}$.

301. $\int x e^{ax} \, dx = \dfrac{e^{ax}}{a^2} (ax - 1), \int x b^{ax} \, dx = \dfrac{x b^{ax}}{a \log b} - \dfrac{b^{ax}}{a^2 (\log b)^2}$.

302. $\int x^2 e^{ax} \, dx = \dfrac{e^{ax}}{a^3} \left(a^2 x^2 - 2ax + 2 \right)$.

303. $\int x^n e^{ax} dx = \frac{1}{a} x^n e^{ax} - \frac{n}{a} \int x^{n-1} e^{ax} dx$, n pos.

304. $\int x^n e^{ax} dx = \frac{e^{ax}}{a^{n+1}} \left[(ax)^n - n(ax)^{n-1} + n(n-1)(ax)^{n-2} - \dots + (-1)^n n! \right]$,

n pos. integ.

305. $\int x^n e^{-ax} dx = -\frac{e^{-ax}}{a^{n+1}} \left[(ax)^n + n(ax)^{n-1} + n(n-1)(ax)^{n-2} + \dots + n! \right]$,

n pos. integ.

306. $\int x^n b^{ax} dx = \frac{x^n b^{ax}}{a \log b} - \frac{n}{a \log b} \int x^{n-1} b^{ax} dx$, n pos.

307. $\int \frac{e^{ax}}{x} dx = \log x + ax + \frac{(ax)^2}{2 \cdot 2!} + \frac{(ax)^3}{3 \cdot 3!} + \dots$

308. $\int \frac{e^{ax}}{x^n} dx = \frac{1}{n-1} \left[-\frac{e^{ax}}{x^{n-1}} + a \int \frac{e^{ax}}{x^{n-1}} dx \right]$, n integ. > 1.

309. $\int \frac{dx}{b + ce^{ax}} = \frac{1}{ab} \left[ax - \log\left(b + ce^{ax} \right) \right]$.

310. $\int \frac{e^{ax} dx}{b + ce^{ax}} = \frac{1}{ac} \log\left(b + ce^{ax} \right)$.

311. $\int \frac{dx}{be^{ax} + ce^{-ax}} = \frac{1}{a\sqrt{bc}} \tan^{-1} \left(e^{ax} \sqrt{\frac{b}{c}} \right)$, b and c pos.

312. $\int e^{ax} \sin bx \, dx = \frac{e^{ax}}{a^2 + b^2} (a \sin bx - b \cos bx)$.

313. $\int e^{ax} \sin bx \sin cx \, dx = \dfrac{e^{ax}[(b-c)\sin(b-c)x + a\cos(b-c)x]}{2\left[a^2 + (b-c)^2\right]}$

$\qquad - \dfrac{e^{ax}[(b+c)\sin(b+c)x + a\cos(b+c)x]}{2\left[a^2 + (b+c)^2\right]}.$

314. $\int e^{ax} \cos bx \, dx = \dfrac{e^{ax}}{a^2 + b^2}(a\cos bx + b\sin bx).$

315. $\int e^{ax} \cos bx \cos cx \, dx = \dfrac{e^{ax}[(b-c)\sin(b-c)x + a\cos(b-c)x]}{2\left[a^2 + (b-c)^2\right]}$

$\qquad + \dfrac{e^{ax}[(b+c)\sin(b+c)x + a\cos(b+c)x]}{2\left[a^2 + (b+c)^2\right]}.$

316. $\int e^{ax} \sin bx \cos cx \, dx = \dfrac{e^{ax}[a\sin(b-c)x - (b-c)\cos(b-c)x]}{2\left[a^2 + (b-c)^2\right]}$

$\qquad + \dfrac{e^{ax}[a\sin(b+c)x - (b+c)\cos(b+c)x]}{2\left[a^2 + (b+c)^2\right]}.$

317. $\int e^{ax} \sin bx \sin(bx + c)dx$

$\qquad = \dfrac{e^{ax}\cos c}{2a} - \dfrac{e^{ax}[a\cos(2bx+c) + 2b\sin(2bx+c)]}{2\left(a^2 + 4b^2\right)}.$

318. $\int e^{ax} \cos bx \cos(bx + c)dx.$

$\qquad = \dfrac{e^{ax}\cos c}{2a} + \dfrac{e^{ax}[a\cos(2bx+c) + 2b\sin(2bx+c)]}{2\left(a^2 + 4b^2\right)}$

319. $\int e^{ax} \sin bx \cos(bx + c)dx.$

$\qquad = \dfrac{e^{ax}\sin c}{2a} + \dfrac{e^{ax}[a\sin(2bx+c) + 2b\cos(2bx+c)]}{2\left(a^2 + 4b^2\right)}$

320. $\int e^{ax} \cos bx \sin(bx + c)dx.$

$$= \frac{e^{ax} \sin c}{2a} + \frac{e^{ax}[a \sin(2bx + c) - 2b \cos(2bx + c)]}{2(a^2 + 4b^2)}$$

321. $\int xe^{ax} \sin bx \, dx = \dfrac{xe^{ax}}{a^2 + b^2}(a \sin bx - b \cos bx)$

$$- \frac{e^{ax}}{\left(a^2 + b^2\right)^2}\left[\left(a^2 - b^2\right)\sin bx - 2ab \cos bx\right].$$

322. $\int xe^{ax} \cos bx \, dx = \dfrac{xe^{ax}}{a^2 + b^2}(a \cos bx + b \sin bx)$

$$- \frac{e^{ax}}{\left(a^2 + b^2\right)^2}\left[\left(a^2 - b^2\right)\cos bx + 2ab \sin bx\right].$$

323. $\int e^{ax} \cos^n bx \, dx = \dfrac{e^{ax} \cos^{n-1} bx(a \cos bx + nb \sin bx)}{a^2 + n^2 b^2}$

$$+ \frac{n(n-1)b^2}{a^2 + n^2 b^2}\int e^{ax} \cos^{n-2} bx \, dx.$$

324. $\int e^{ax} \sin^n bx \, dx = \dfrac{e^{ax} \sin^{n-1} bx(a \sin bx - nb \cos bx)}{a^2 + n^2 b^2}$

$$+ \frac{n(n-1)b^2}{a^2 + n^2 b^2}\int e^{ax} \sin^{n-2} bx \, dx.$$

325. $\int \log ax \, dx = x \log ax - x.$

326. $\int x \log ax \, dx = \dfrac{x^2}{2} \log ax - \dfrac{x^2}{4}.$

327. $\int x^2 \log ax \, dx = \dfrac{x^3}{3} \log ax - \dfrac{x^3}{9}.$

328. $\int (\log ax)^2 dx = x(\log ax)^2 - 2x \log ax + 2x.$

329. $\int (\log ax)^n dx = x(\log ax)^n - n \int (\log ax)^{n-1} dx, n$ pos.

330. $\int x^n \log ax \, dx = x^{n+1} \left[\dfrac{\log ax}{n+1} - \dfrac{1}{(n+1)^2} \right], n \neq -1.$

331. $\int x^n (\log ax)^m \, dx = \dfrac{x^{n+1}}{n+1} (\log ax)^m - \dfrac{m}{n+1} \int x^n (\log ax)^{m-1} dx.$

332. $\int \dfrac{(\log ax)^n}{x} dx = \dfrac{(\log ax)^{n+1}}{n+1}, n \neq -1.$

333. $\int \dfrac{dx}{x \log ax} = \log(\log ax).$

334. $\int \dfrac{dx}{x(\log ax)^n} = -\dfrac{1}{(n-1)(\log ax)^{n-1}}.$

335. $\int \dfrac{x^n dx}{(\log ax)^m} = \dfrac{-x^{n+1}}{(m-1)(\log ax)^{m-1}} + \dfrac{n+1}{m-1} \int \dfrac{x^n dx}{(\log ax)^{m-1}}, m \neq 1.$

336. $\int \dfrac{x^n dx}{\log ax} = \dfrac{1}{a^{n+1}} \int \dfrac{e^y dy}{y}, y = (n+1) \log ax.$

337. $\int \dfrac{x^n dx}{\log ax} = \dfrac{1}{a^{n+1}} \left[\log|\log ax| + (n+1) \log ax + \dfrac{(n+1)^2 (\log ax)^2}{2 \cdot 2!} \right.$
$\left. + \dfrac{(n+1)^3 (\log ax)^3}{3 \cdot 3!} + ... \right].$

338. $\int \dfrac{dx}{\log ax} = \dfrac{1}{a} \left[\log|\log ax| + \log ax + \dfrac{(\log ax)^2}{2 \cdot 2!} + \dfrac{(\log ax)^3}{3 \cdot 3!} + ... \right].$

339. $\int \sin(\log ax)dx = \dfrac{x}{2}[\sin(\log ax) - \cos(\log ax)].$

340. $\int \cos(\log ax)dx = \dfrac{x}{2}[\sin(\log ax) + \cos(\log ax)].$

341. $\int e^{ax} \log bx\, dx = \dfrac{1}{a} e^{ax} \log bx - \dfrac{1}{a}\int \dfrac{e^{ax}}{x}\, dx.$

Expressions Containing Inverse Trigonometric Functions

342. $\int \sin^{-1} ax\, dx = x \sin^{-1} ax + \dfrac{1}{a}\sqrt{1 - a^2 x^2}.$

343. $\int \left(\sin^{-1} ax\right)^2 dx = x \left(\sin^{-1} ax\right)^2 - 2x + \dfrac{2}{a}\sqrt{1 - a^2 x^2}\, \sin^{-1} ax.$

344. $\int x \sin^{-1} ax\, dx = \dfrac{x^2}{2} \sin^{-1} ax - \dfrac{1}{4a^2} \sin^{-1} ax + \dfrac{x}{4a}\sqrt{1 - a^2 x^2}.$

345. $\int x^n \sin^{-1} ax\, dx = \dfrac{x^{n+1}}{n+1} \sin^{-1} ax - \dfrac{a}{n+1}\int \dfrac{x^{n+1}dx}{\sqrt{1 - a^2 x^2}}, n \neq -1.$

346. $\int \dfrac{\sin^{-1} ax\, dx}{x} = ax + \dfrac{1}{2 \cdot 3 \cdot 3}(ax)^3 + \dfrac{1 \cdot 3}{2 \cdot 4 \cdot 5 \cdot 5}(ax)^5$

$\qquad\qquad + \dfrac{1 \cdot 3 \cdot 5}{2 \cdot 4 \cdot 6 \cdot 7 \cdot 7}(ax)^7 + ..., a^2 x^2 < 1.$

347. $\int \dfrac{\sin^{-1} ax\, dx}{x^2} = -\dfrac{1}{x} \sin^{-1} ax - a \log\left|\dfrac{1 + \sqrt{1 - a^2 x^2}}{ax}\right|.$

348. $\int \cos^{-1} ax\, dx = x \cos^{-1} ax - \dfrac{1}{a}\sqrt{1 - a^2 x^2}.$

349. $\int \left(\cos^{-1} ax\right)^2 dx = x\left(\cos^{-1} ax\right)^2 - 2x - \dfrac{2}{a}\sqrt{1 - a^2 x^2}\,\cos^{-1} ax.$

350. $\int x\cos^{-1} ax\,dx = \dfrac{x^2}{2}\cos^{-1} ax - \dfrac{1}{4a^2}\cos^{-1} ax - \dfrac{x}{4a}\sqrt{1 - a^2 x^2}.$

351. $\int x^n \cos^{-1} ax\,dx = \dfrac{x^{n+1}}{n+1}\cos^{-1} ax + \dfrac{a}{n+1}\int\dfrac{x^{n+1}dx}{\sqrt{1 - a^2 x^2}}, \; n \neq -1.$

352. $\int \dfrac{\cos^{-1} ax\,dx}{x} = \dfrac{\pi}{2}\log|ax| - ax - \dfrac{1}{2\cdot3\cdot3}(ax)^3 - \dfrac{1\cdot3}{2\cdot4\cdot5\cdot5}(ax)^5$

$$- \dfrac{1\cdot3\cdot5}{2\cdot4\cdot6\cdot7\cdot7}(ax)^7 - ..., a^2 x^2 < 1.$$

353. $\int \dfrac{\cos^{-1} ax\,dx}{x^2} = -\dfrac{1}{x}\cos^{-1} ax + a\log\left|\dfrac{1 + \sqrt{1 - a^2 x^2}}{ax}\right|.$

354. $\int \tan^{-1} ax\,dx = x\tan^{-1} ax - \dfrac{1}{2a}\log\left(1 + a^2 x^2\right).$

355. $\int x^n \tan^{-1} ax\,dx = \dfrac{x^{n+1}}{n+1}\tan^{-1} ax - \dfrac{a}{n+1}\int\dfrac{x^{n+1}dx}{1 + a^2 x^2}, \; n \neq -1.$

356. $\int \dfrac{\tan^{-1} ax\,dx}{x^2} = -\dfrac{1}{x}\tan^{-1} ax - \dfrac{a}{2}\log\left(\dfrac{1 + a^2 x^2}{a^2 x^2}\right).$

357. $\int \cot^{-1} ax\,dx = x\cot^{-1} ax + \dfrac{1}{2a}\log\left(1 + a^2 x^2\right).$

358. $\int x^n \cot^{-1} ax\,dx = \dfrac{x^{n+1}}{n+1}\cot^{-1} ax + \dfrac{a}{n+1}\int\dfrac{x^{n+1}dx}{1 + a^2 x^2}, \; n \neq -1.$

359. $\int \dfrac{\cot^{-1} ax \, dx}{x^2} = -\dfrac{1}{x} \cot^{-1} ax + \dfrac{a}{2} \log\left(\dfrac{1 + a^2 x^2}{a^2 x^2} \right).$

360. $\int \sec^{-1} ax \, dx = x \sec^{-1} ax - \dfrac{1}{a} \log\left(ax + \sqrt{a^2 x^2 - 1} \right).$

361. $\int x^n \sec^{-1} ax \, dx = \dfrac{x^{n+1}}{n+1} \sec^{-1} ax \pm \dfrac{1}{n+1} \int \dfrac{x^n \, dx}{\sqrt{a^2 x^2 - 1}}, n \neq -1.$

Use + sign when $\dfrac{\pi}{2} < \sec^{-1} ax < \pi$; sign when $0 < \sec^{-1} ax < \dfrac{\pi}{2}.$

362. $\int \mathrm{cosec}^{-1} ax \, dx = x \, \mathrm{cosec}^{-1} ax + \dfrac{1}{a} \log\left(ax + \sqrt{a^2 x^2 - 1} \right).$

363. $\int x^n \, \mathrm{cosec}^{-1} ax \, dx = \dfrac{x^{n+1}}{n+1} \mathrm{cosec}^{-1} ax \pm \dfrac{1}{n+1} \int \dfrac{x^n \, dx}{\sqrt{a^2 x^2 - 1}}, n \neq -1.$

Use + sign when $0 < \mathrm{cosec}^{-1} ax < \dfrac{\pi}{2}$; − sign when $-\dfrac{\pi}{2} < \mathrm{cosec}^{-1} ax < 0.$

Definite Integrals

364. $\displaystyle\int_0^\infty \dfrac{a \, dx}{a^2 + x^2} = \dfrac{\pi}{2},$ if $a > 0$: 0, if $a = 0$; $\dfrac{-\pi}{2},$ if $a < 0.$

365. $\displaystyle\int_0^\infty x^{n-1} e^{-x} dx = \int_0^1 \left[\log_e (1 / x) \right]^{n-1} dx = \Gamma(n).$

$\Gamma(n + 1) = n \cdot \Gamma(n),$ if $n > 0.$ $\Gamma(2) = \Gamma(1) = 1.$

$\Gamma(n + 1) = n!,$ if n is an integer. $\Gamma\left(\dfrac{1}{2}\right) = \sqrt{\pi}.$

$\Gamma(n) = \Pi(n - 1).$ $Z(y) = D_y \left[\log_e \Gamma(y) \right].$

$Z(1) = -0.5772157\ldots$(See integral 418).

366. $\int_0^\infty e^{-zx} \cdot z^n \cdot x^{n-1}\,dx = \Gamma(n), \quad z > 0.$

367. $\int_0^1 x^{m-1}(1-x)^{n-1}\,dx = \int_0^\infty \dfrac{x^{m-1}dx}{(1+x)^{m+n}} = \dfrac{\Gamma(m)\Gamma(n)}{\Gamma(m+n)}.$

368. $\int_0^\infty \dfrac{x^{n-1}}{1+x}\,dx = \dfrac{\pi}{\sin n\pi}, \; 0 < n < 1.$

369. $\int_0^{\frac{\pi}{2}} \sin^n x\,dx = \int_0^{\frac{\pi}{2}} \cos^n x\,dx$

$$= \frac{1}{2}\sqrt{\pi} \cdot \frac{\Gamma\left(\dfrac{n}{2}+\dfrac{1}{2}\right)}{\Gamma\left(\dfrac{n}{2}+1\right)}, \text{if } n > -1;$$

$$= \frac{1\cdot3\cdot5...(n-1)}{2\cdot4\cdot6...(n)}\cdot\frac{\pi}{2}, \text{if } n \text{ is an even integer;}$$

$$= \frac{2\cdot4\cdot6...(n-1)}{1\cdot3\cdot5\cdot7...n}, \text{if } n \text{ is an odd integer.}$$

370. $\int_0^\infty \dfrac{\sin^2 x}{x^2}\,dx = \dfrac{\pi}{2}.$

371. $\int_0^\infty \dfrac{\sin ax}{x}\,dx = \dfrac{\pi}{2}, \text{if } a > 0.$

372. $\int_0^\infty \dfrac{\sin x \cos ax}{x}\,dx \quad = \quad 0, \text{if } a < -1, \quad or \quad a > 1;$

$$= \quad \frac{\pi}{4}, \text{if } a = -1, \quad or \quad a = 1;$$

$$= \quad \frac{\pi}{2}, \text{if } -1 < a < 1.$$

373. $\int_0^\pi \sin^2 ax\,dx = \int_0^\pi \cos^2 ax\,dx = \dfrac{\pi}{2}.$

374. $\int_0^{\pi/a} \sin ax \cdot \cos ax \, dx = \int_0^\pi \sin ax \cdot \cos ax \, dx = 0.$

375. $\int_0^\pi \sin ax \sin bx \, dx = \int_0^\pi \cos ax \cos bx \, dx = 0, a \neq b.$

376. $\int_0^\pi \sin ax \cos bx \, dx = \dfrac{2a}{a^2 - b^2}$, if $a - b$ is odd;

$\qquad\qquad\qquad = 0, \qquad$ if $a - b$ is even.

377. $\int_0^\infty \dfrac{\sin ax \sin bx}{x^2} dx = \dfrac{1}{2} \pi a$, if $a < b.$

378. $\int_0^\infty \cos\left(x^2\right) dx = \int_0^\infty \sin\left(x^2\right) dx = \dfrac{1}{2} \sqrt{\dfrac{\pi}{2}}.$

379. $\int_0^\infty e^{-a^2 x^2} dx = \dfrac{\sqrt{\pi}}{2a} = \dfrac{1}{2a} \Gamma\left(\dfrac{1}{2}\right)$, if $a > 0.$

380. $\int_0^\infty x^n \cdot e^{-ax} \, dx = \dfrac{\Gamma(n+1)}{a^{n+1}},$

$\qquad\qquad\qquad = \dfrac{n!}{a^{n+1}}$, if n is a positive integer, $a > 0.$

381. $\int_0^\infty x^{2n} e^{-ax^2} dx = \dfrac{1 \cdot 3 \cdot 5 \ldots (2n-1)}{2^{n+1} a^n} \sqrt{\dfrac{\pi}{a}}.$

382. $\int_0^\infty \sqrt{x} e^{-ax} dx = \dfrac{1}{2a} \sqrt{\dfrac{\pi}{2}}.$

383. $\int_0^\infty \dfrac{e^{-ax}}{\sqrt{x}} dx = \sqrt{\dfrac{\pi}{a}}.$

384. $\int_0^\infty e^{\left(-x^2 - a^2/x^2\right)} dx = \dfrac{1}{2} e^{-2a} \sqrt{\pi}$, if $a > 0.$

385. $\int_0^\infty e^{-ax} \cos bx \, dx = \dfrac{a}{a^2 + b^2}$, if $a > 0$.

386. $\int_0^\infty e^{-ax} \sin bx \, dx = \dfrac{b}{a^2 + b^2}$, if $a > 0$.

387. $\int_0^\infty \dfrac{e^{-ax} \sin x}{x} \, dx = \operatorname{ctn}^{-1} a, \ a > 0$.

388. $\int_0^\infty e^{-a^2 x^2} \cos bx \, dx = -\dfrac{\sqrt{\pi} \cdot e^{-b^2/4a^2}}{2a}$, if $a > 0$.

389. $\int_0^1 (\log x)^n \, dx = (-1)^n . n!, \ n$ pos. integ.

390. $\int_0^1 \dfrac{\log x}{1-x} \, dx = -\dfrac{\pi^2}{6}$.

391. $\int_0^1 \dfrac{\log x}{1+x} \, dx = -\dfrac{\pi^2}{12}$.

392. $\int_0^1 \dfrac{\log x}{\sqrt{1-x^2}} \, dx = -\dfrac{\pi^2}{8}$.

393. $\int_0^1 \dfrac{\log x}{\sqrt{1-x^2}} \, dx = -\dfrac{\pi}{2} \log 2$.

394. $\int_0^1 \log\left(\dfrac{1+x}{1-x}\right) . \dfrac{dx}{x} = \dfrac{\pi^2}{4}$.

395. $\int_0^\infty \log\left(\dfrac{e^x + 1}{e^x - 1}\right) dx = \dfrac{\pi^2}{4}$.

396. $\int_0^1 \dfrac{dx}{\sqrt{\log(1/x)}} = \sqrt{\pi}.$

397. $\int_0^1 \log|\log x| \, dx = \int_0^\infty e^{-x} \log x \, dx = -\gamma = -0.577\ 2157....$

398. $\int_0^{\frac{\pi}{2}} \log \sin x \, dx = \int_0^{\frac{\pi}{2}} \log \cos x \, dx = -\dfrac{\pi}{2} \log_e 2.$

399. $\int_0^\pi x \log \sin x \, dx = -\dfrac{\pi^2}{2} \log_e 2.$

400. $\int_0^1 \left(\log \dfrac{1}{x} \right)^{\frac{1}{2}} dx = \dfrac{\sqrt{\pi}}{2}.$

401. $\int_0^1 \left(\log \dfrac{1}{x} \right)^{-\frac{1}{2}} dx = \sqrt{\pi}.$

402. $\int_0^1 x^m \left(\log \dfrac{1}{x} \right)^n dx = \dfrac{\Gamma(n+1)}{(m+1)^{n+1}},$ if $m + 1 > 0, \; n + 1 > 0.$

403. $\int_0^\pi \log(a \pm b \cos x) \, dx = \pi \log \left(\dfrac{a + \sqrt{a^2 - b^2}}{2} \right), \; a \underset{=}{>} b.$

404. $\int_0^\pi \dfrac{\log(1 + \sin a \cos x)}{\cos x} \, dx = \pi a.$

405. $\int_0^1 \dfrac{x^b - x^a}{\log x} \, dx = \log \dfrac{1+b}{1+a}.$

406. $\displaystyle\int_0^\pi \frac{dx}{a + b\cos x} = \frac{\pi}{\sqrt{a^2 - b^2}}$, if $a > b > 0$.

407. $\displaystyle\int_0^{\frac{\pi}{2}} \frac{dx}{a + b\cos x} = \frac{\cos^{-1}\left(\dfrac{b}{a}\right)}{\sqrt{a^2 - b^2}}$, $a > b$.

408. $\displaystyle\int_0^\infty \frac{\cos ax\, dx}{1 + x^2} = \frac{\pi}{2}e^{-a}$, if $a > 0$; $= \frac{\pi}{2}e^a$, if $a < 0$.

409. $\displaystyle\int_0^\infty \frac{\cos x dx}{\sqrt{x}} = \int_0^\infty \frac{\sin x dx}{\sqrt{x}} = \sqrt{\frac{\pi}{2}}$.

410. $\displaystyle\int_0^\infty \frac{e^{-ax} - e^{-bx}}{x}\, dx = \log\frac{b}{a}$.

411. $\displaystyle\int_0^\infty \frac{\tan^{-1} ax - \tan^{-1} bx}{x}\, dx = \frac{\pi}{2}\log\frac{a}{b}$.

412. $\displaystyle\int_0^\infty \frac{\cos ax - \cos bx}{x}\, dx = \log\frac{b}{a}$.

413. $\displaystyle\int_0^{\frac{\pi}{2}} \frac{dx}{a^2 \cos^2 x + b^2 \sin^2 x} = \frac{\pi}{2ab}$.

414. $\displaystyle\int_0^{\frac{\pi}{2}} \frac{dx}{\left(a^2 \cos^2 x + b^2 \sin^2 x\right)^2} = \frac{\pi\left(a^2 + b^2\right)}{4a^3 b^3}$.

415. $\displaystyle\int_0^\pi \frac{(a - b\cos x)dx}{a^2 - 2ab\cos x + b^2}$ $= 0$, if $a^2 < b^2$;

$\qquad\qquad\qquad = \dfrac{\pi}{a}$, if $a^2 > b^2$;

$\qquad\qquad\qquad = \dfrac{\pi}{2a}$, if $a = b$.

416. $\int\limits_{0}^{1}\dfrac{1+x^2}{1+x^4}dx=\dfrac{\pi}{4}\sqrt{2}.$

417. $\int\limits_{0}^{1}\dfrac{\log(1+x)}{x}dx=\dfrac{1}{1^2}-\dfrac{1}{2^2}+\dfrac{1}{3^2}-\dfrac{1}{4^2}+...=\dfrac{\pi^2}{12}.$

418. $\int\limits_{+\infty}^{1}\dfrac{e^{-xu}}{u}du=\gamma+\log x-x+\dfrac{x^2}{2\cdot 2!}-\dfrac{x^3}{3\cdot 3!}+\dfrac{x^4}{4\cdot 4!}-...,$ where

$\gamma=\lim\limits_{t\to\infty}\left(1+\dfrac{1}{2}+\dfrac{1}{3}+...+\dfrac{1}{t}-\log t\right)=0.5772157...,0<x<\infty.$

419. $\int\limits_{+\infty}^{1}\dfrac{\cos xu}{u}du=\gamma+\log x-\dfrac{x^2}{2\cdot 2!}+\dfrac{x^4}{4\cdot 4!}-\dfrac{x^6}{6\cdot 6!}+...,$

where $\gamma=0.5772157...,0<x<\infty.$

420. $\int\limits_{0}^{1}\dfrac{e^{xu}-e^{-xu}}{u}du=2\left(x+\dfrac{x^3}{3\cdot 3!}+\dfrac{x^5}{5\cdot 5!}+...\right),0<x<\infty.$

421. $\int\limits_{0}^{1}\dfrac{1-e^{-xu}}{u}du=x-\dfrac{x^2}{2\cdot 2!}+\dfrac{x^3}{3\cdot 3!}-\dfrac{x^4}{4\cdot 4!}+...,0<x<\infty.$

422. $\int\limits_{0}^{\frac{\pi}{2}}\dfrac{dx}{\sqrt{1-K^2\sin^2 x}}=\dfrac{\pi}{2}\left[1+\left(\dfrac{1}{2}\right)^2 K^2+\left(\dfrac{1\cdot 3}{2\cdot 4}\right)^2 K^4\right.$

$\left.+\left(\dfrac{1\cdot 3\cdot 5}{2\cdot 4\cdot 6}\right)^2 K^6+...\right],$ if $K^2<1.$

423. $\int\limits_{0}^{\frac{\pi}{2}}\sqrt{1-K^2\sin^2 x}\,dx=\dfrac{\pi}{2}\left[1-\left(\dfrac{1}{2}\right)^2 K^2-\left(\dfrac{1\cdot 3}{2\cdot 4}\right)^2\dfrac{K^4}{3}\right.$

$\left.-\left(\dfrac{1\cdot 3\cdot 5}{2\cdot 4\cdot 6}\right)^2\dfrac{K^6}{5}-...\right],$ if $K^2<1.$

424. $f(x) = \dfrac{1}{2}a_0 + a_1 \cos\dfrac{\pi x}{c} + a_2 \cos\dfrac{2\pi x}{c} + ... + b_1 \sin\dfrac{\pi x}{c}$

$$+ b_2 \sin\dfrac{2\pi x}{c} + ..., -c < x < +c,$$

where $a_m = \dfrac{1}{c}\displaystyle\int_{-c}^{+c} f(x)\cos\dfrac{m\pi x}{c}\,dx,\ b_m = \dfrac{1}{c}\displaystyle\int_{-c}^{+c} f(x)\sin\dfrac{m\pi x}{c}\,dx.$

(*Fourier Series*)

425. $\displaystyle\int_0^\infty e^{-ax}\cosh bx\,dx = \dfrac{a}{a^2 - b^2},\ 0 \leq |b| < a.$

426. $\displaystyle\int_0^\infty e^{-ax}\sinh bx\,dx = \dfrac{b}{a^2 - b^2},\ 0 \leq |b| < a.$

427. $\displaystyle\int_0^\infty xe^{-ax}\sin bx\,dx = \dfrac{2ab}{\left(a^2 + b^2\right)^2},\ a > 0.$

428. $\displaystyle\int_0^\infty xe^{-ax}\cos bx\,dx = \dfrac{a^2 - b^2}{\left(a^2 + b^2\right)^2},\ a > 0.$

429. $\displaystyle\int_0^\infty x^2 e^{-ax}\sin bx\,dx = \dfrac{2b\left(3a^2 - b^2\right)}{\left(a^2 + b^2\right)^3},\ a > 0.$

430. $\displaystyle\int_0^\infty x^2 e^{-ax}\cos bx\,dx = \dfrac{2a\left(a^2 - 3b^2\right)}{\left(a^2 + b^2\right)^3},\ a > 0.$

431. $\displaystyle\int_0^\infty x^3 e^{-ax}\sin bx\,dx = \dfrac{24ab\left(a^2 - b^2\right)}{\left(a^2 + b^2\right)^4},\ a > 0.$

432. $\displaystyle\int_0^\infty x^3 e^{-ax}\cos bx\,dx = \dfrac{6\left(a^4 - 6a^2 b^2 + b^4\right)}{\left(a^2 + b^2\right)^4},\ a > 0.$

433. $\displaystyle\int_0^\infty x^n e^{-ax} \sin bx \, dx = \frac{i \cdot n! \left[(a-ib)^{n+1} - (a+ib)^{n+1} \right]}{2 \left(a^2 + b^2 \right)^{n+1}}, \, a > 0.$

434. $\displaystyle\int_0^\infty x^n e^{-ax} \cos bx \, dx = \frac{n! \left[(a-ib)^{n+1} + (a+ib)^{n+1} \right]}{2 \left(a^2 + b^2 \right)^{n+1}}, \, a > 0.$

435. $\displaystyle\int_0^\infty e^{-x} \log x \, dx = -\gamma = -0.577\,2157\dots$ (Also See Art. 418.)

436. $\displaystyle\int_0^\infty \left(\frac{1}{1-e^{-x}} - \frac{1}{x} \right) e^{-x} dx = \gamma = 0.577\,2157\dots$

437. $\displaystyle\int_0^\infty \frac{1}{x} \left(\frac{1}{1+x} - e^{-x} \right) dx = \gamma = 0.577\,2157\dots$

438. $\displaystyle\int_0^1 \frac{\left(1 - e^{-x} - e^{-\frac{1}{x}} \right)}{x} dx = \gamma = 0.577\,2157\dots$

8

VECTORS

150 DEFINITIONS: ANALYTIC REPRESENTATION

A vector V is a quantity which is completely specified by a magnitude and a direction. A vector may be represented geometrically by a directed line segment $V = \overrightarrow{OA}$. A scalar S is a quantity which is completely specified by a magnitude.

Let $i, j,$ and k represent three vectors of unit magnitude along the three mutually perpendicular lines $OX,$ $OY,$ and $OZ,$ respectively. Let V be a vector in space, and $a, b,$ and c the magnitudes of the projections of V along the three lines $OX,$ $OY,$ and $OZ,$ respectively. Then V may be represented by $V = ai + bj + ck$. The magnitude of V is $|V| = +\sqrt{a^2 + b^2 + c^2}$, and the direction cosines of V are such that $\cos \alpha : \cos \beta : \cos \gamma = a : b : c$.

151 VECTOR SUM V OF n VECTORS

Let $V_1, V_2, \dots V_n$ be n vectors given by $V_1 = a_1 i + b_1 j + c_1 k$, etc. Then the *sum* is $V = V_1 + V_2 + \dots + V_n = (a_1 + a_2 + \dots + a_n) i + (b_1 + b_2 + \dots + bn) j + (c_1 + c_2 + \dots + c_n) k$.

152 PRODUCT OF A SCALAR S AND A VECTOR V

$$SV = (Sa)\, i + (Sb)\, j + (Sc)\, k.$$

$$(S_1 + S_2)V = S_1 V + S_2 V \cdot (V_1 + V_2)S = V_1 S + V_2 S.$$

153   SCALAR PRODUCT OF TWO VECTORS: $V_1.V_2$.

$V_1 \cdot V_2 = |V_1| |V_2| \cos \phi,$ where ϕ is the angle from V_1 to V_2.

$V_1 \cdot V_2 = a_1 a_2 + b_1 b_2 + c_1 c_2 = V_2 \cdot V_1$ $V_1 \cdot V_1 = |V_1|^2.$

$(V_1 + V_2) \cdot V_3 = V_1 \cdot V_3 + V_2 \cdot V_3.$

$V_1 \cdot (V_2 + V_3) = V_1 \cdot V_2 + V_1 \cdot V_3.$

$i \cdot i = j \cdot j = k \cdot k = 1.$ $i \cdot j = j \cdot k = k \cdot i = 0.$

154   VECTORS PRODUCT OF TWO VECTORS: $V_1 \times V_2$.

$V_1 \times V_2 = |V_1| |V_2| \sin \phi \, \mathbf{1}$, where ϕ is the angle from V_1 to V_2 and $\mathbf{1}$ is a unit vector perpendicular to the plane of V_1 and V_2 and so directed that a right-handed screw driven in the direction of $\mathbf{1}$ would carry V_1 into V_2.

$$V_1 \times V_2 = -V_2 \times V_1 = (b_1 c_2 - b_2 c_1) i + (c_1 a_2 - c_2 a_1) j$$
$$+ (a_1 b_2 - a_2 b_1) k.$$

$(V_1 + V_2) \times V_3 = V_1 \times V_3 + V_2 \times V_3.$

$V_1 \times (V_2 + V_3) = V_1 \times V_2 + V_1 \times V_3.$

$V_1 \times (V_2 \times V_3) = V_2(V_1 \cdot V_3) - V_3(V_1 \cdot V_2).$

$i \times i = j \times j = k \times k = 0, \quad i \times j = k, \quad j \times k = i, k \times i = j,$

$$V_1 \cdot (V_2 \times V_3) = (V_1 \times V_2) \cdot V_3 = V_2 \cdot (V_3 \times V_1)$$

$$= [V_1 \, V_2 \, V_3] = \begin{vmatrix} a_1 & a_2 & a_3 \\ b_1 & b_2 & b_3 \\ c_1 & c_2 & c_3 \end{vmatrix}.$$

155   DIFFERENTIATION OF VECTORS

$V = ai + bj + ck.$ If $V_1, V_2 \ldots..$ are functions of a scalar variable t, then

$$\frac{d}{dt}(V_1 + V_2 + \ldots) = \frac{dV_1}{dt} + \frac{dV_2}{dt} + \ldots,$$

where
$$\frac{dV_1}{dt} = \frac{da_1}{dt}\boldsymbol{i} + \frac{db_1}{dt}\boldsymbol{j} + \frac{dc_1}{dt}\boldsymbol{k}, \text{ etc.}$$

$$\frac{d}{dt}(V_1 \cdot V_2) = \frac{dV_1}{dt} \cdot V_2 + V_2 \cdot \frac{dV_2}{dt}.$$

$$\frac{d}{dt}(V_1 \times V_2) = \frac{dV_1}{dt} \times V_2 + V_1 \times \frac{dV_2}{dt}.$$

$$V \cdot \frac{dV}{dt} = |V|\frac{d|V|}{dt}. \text{ If } |V| \text{ is constant, } V \cdot \frac{dV}{dt} = 0.$$

$$\text{grad } S \equiv \nabla S \equiv \frac{\partial S}{\partial x}\boldsymbol{i} + \frac{\partial S}{\partial y}\boldsymbol{j} + \frac{\partial S}{\partial z}\boldsymbol{k}, \text{ where S is a scalar.}$$

$$\text{div } V \equiv \nabla \cdot V \equiv \frac{\partial a}{\partial x} + \frac{\partial b}{\partial y} + \frac{\partial c}{\partial z}. \text{ (divergence of } \mathbf{V}).$$

$$\text{curl } V \equiv \text{rot } V \equiv \begin{vmatrix} i & j & k \\ \dfrac{\partial}{\partial x} & \dfrac{\partial}{\partial y} & \dfrac{\partial}{\partial z} \\ a & b & c \end{vmatrix} \equiv \nabla \times V.$$

$$\text{div grad } S \equiv \nabla^2 S \equiv \frac{\partial^2 S}{\partial x^2} + \frac{\partial^2 S}{\partial y^2} + \frac{\partial^2 S}{\partial z^2}.$$

$$\nabla^2 V \equiv \boldsymbol{i}\nabla^2 a + \boldsymbol{j}\nabla^2 b + \boldsymbol{k}\nabla^2 c.$$

curl grad $S = 0$. div curl $V = 0$.

curl curl V = grad div V $- \nabla^2 V$.

156   GREEN'S THEOREM

Let $\boldsymbol{F}$ be a vector and V be a volume bounded by a surface S. Then

$$\iiint_{(V)} \operatorname{div} \boldsymbol{F} \, dV = \iiint_{(V)} \nabla \cdot \boldsymbol{F} \, dV = \iint_{(S)} F \cdot d\boldsymbol{S},$$

where the integrations are to be carried out over the volume V and the surface S.

157   STOKES' THEOREM

Let $\boldsymbol{F}$ be a vector and $dr = dx \, \boldsymbol{i} + dy \, \boldsymbol{j} + dz \, \boldsymbol{k}$, and S be a surface bounced by a simple closed curve C. Then

$$\int_{(C)} \boldsymbol{F} \cdot d\boldsymbol{r} = \iint_{(S)} \operatorname{curl} \boldsymbol{F} \cdot d\boldsymbol{S} = \iint_{(S)} \nabla \times \boldsymbol{F} \cdot d\boldsymbol{S}.$$

PART II

SCIENTIFIC DATA FOR ENGINEERS

Constants

e = Base of natural logarithms ≈ 2.71828

$\log_{10} e \approx 0.434294$ $\log_e 10 \approx 2.30259$

$\log_{10} N \approx \log_e N \times 0.4343 \log_e N \approx \log_{10} N \times 2.3026$

1-radian ≈ 57° · 2958 ≈ 57° 17′ 45″ $\pi \approx 3 \cdot 14159265$

$\log_{10} \pi \approx 0.49715$ $\dfrac{1}{\pi} \approx 0 \cdot 31831$ $\dfrac{\pi}{180} \approx 0 \cdot 01745$ $\pi^3 \approx 9 \cdot 8696$

1

SI UNITS

When making measurements of a physical quantity, the final result is expressed as a number followed by the unit. The number expresses the ratio of the measured quantity to some fixed standard and the unit is the name or symbol for the standard. Over the years, a large number of standards have been defined for physical measurement and many systems of units have evolved. Recently, there has been an attempt to simplify the language of science by the adoption of a system of units, the System International Units, which is intended to be used universally. This system of units, SI, was the outcome of a resolution of the 9th General Conference of Weights and Measures (CGPM) in 1948, which instructed an international committee to "study the establishment of a complete set of rules for units of measurement." The constants in this book are given in SI except where stated otherwise.

SI contains three classes of units: (*i*) base units, (*ii*) derived units, and (*iii*) supplementary units.

Base Units: There are seven base units:

(i)   the meter, the standard of length,

(ii)   the kilogram, the standard of mass,

(iii)   the second, the standard of time,

(iv)   the ampere, the standard of electric current,

(v)   the kelvin, the standard of temperature,

(vi) the candela, the standard of luminous intensity, and

(vii) the mole, the standard of amount of substance.

Derived Units: Derived units can be formed by combining base units. Thus the unit of force can be produced by combining the first three base units. Often derived units are given names, for example, the unit of force is the *newton*.

Supplementary Units: Two supplementary units are at present defined, the radian and the steradian, which are the units for plane and solid angles, respectively.

SI Prefixes and Multiplication Factors: To obtain multiples and submultiples of units, standard prefixes are used as shown below:

Multiplication factor	Symbol	Prefix
$1\,000\,000\,000\,000 = 10^{12}$	T	tera
$1\,000\,000\,000 = 10^{9}$	G	giga
$1\,000\,000 = 10^{6}$	M	mega
$1\,000 = 10^{3}$	k	kilo
$100 = 10^{2}$	h	hecto
$10 = 10^{1}$	da	deca
$0.1 = 10^{-1}$	d	deci
$0.01 = 10^{-2}$	c	centi
$0.001 = 10^{-3}$	m	milli
$0.000\,001 = 10^{-6}$	µ	micro
$0.000\,000\,001 = 10^{-9}$	n	nano
$0.000\,000\,000\,001 = 10^{-12}$	p	pico
$0.000\,000\,000\,000\,001 = 10^{-15}$	f	femto
$0.000\,000\,000\,000\,000\,001 = 10^{-18}$	a	atto

It should be noted that masses are still expressed as multiples of the gram, although the base unit is the kilogram. Thus 10^{-6} kg should be written as 1 mg.

Other systems of units: Some other systems of units are still in common use. Thus for mechanical measurements, the British or fps system is still largely used, while for electrical measurements, the electrostatic and electromagnetic cgs systems are by no means obsolete. In the pages which follow, these systems of units are discussed and tables are included to help in conversion from one system to another.

2

THE FUNDAMENTAL MECHANICAL UNITS

(A) SI UNITS

In any system of measurement in mechanics, three fundamental units are required. These are the units of length, mass, and time. The base units as used in SI are the meter, kilogram, and second.

The meter (m)

This is defined as 1 650 763-73 of the wavelength, in vacuo of the orange light emitted by $^{86}_{38}\mathrm{Kr}$ in the transition $2\mathrm{p}_{10}$ to $5\mathrm{d}_{5}$.

The kilogram (kg)

This is defined as the mass of a platinum-iridium cylinder kept at Sévres. Originally intended to be the mass of a cubic decimeter of water at its maximum density, the cylinder was subsequently found to be 28 parts per million too large. The cylinder was then taken as an arbitrary standard of mass, while the volume of water which had the same mass (at maximum density) was defined to be 1 liter (l). Thus 1 litre = 1000.028 cm^3. The 1964 General Conference on Weights and Measures redefined the liter to be a cubic decimeter but recommended that this unit should not be used in work of high precision.

The second (s)

This is the time taken by 9 192 631 770 cycles of the radiation from the hyperfine transition in cesium when unperturbed by external fields. Alternatively the *ephemeris* second is defined as 1/31 556 925-974 7 of the tropical year for 1900.

Derived units of length, mass, and time

Through common usage, certain multiples and submultiples of the three fundamental units have been given names. A list of the more common ones is given below as they have been in frequent use. None of them is a recognized SI unit.

Length and area	Mass	Time
Micron (μm) = 10^{-6} m	Tonne (t) = 10^6 g	Minute (min) = 60 s
Angstrom (Å) = 10^{-10} m	= 1000 kg	Hour (h) = 3 600 s
Fermi (fm) = 10^{-15} m		Day (d) = 86 400 s
Are (*a*) = 100 m²		Year (a) $\simeq 3.1557 \times 10^7$ s
Barn (*b*) = 10^{-28} m²		

The radian (rad) is the plane angle between two radii of a circle that cut off on the circumference of an arc equal in length to the radius.

The steradian (sr) is the solid angle which, having its vertex in the center of a sphere, cuts off an area of the surface of the sphere equal to that of a square with sides of length equal to the radius of the sphere.

Other units of angular measures are:

The degree (°) is a unit of angle equal to ($\pi/180$) rad.
The minute of arc (′) is (1/60) degree and thus is equal to ($\pi/10\ 800$) rad.
The second of arc (″) is (1/60) minute and thus is equal to ($\pi/648\ 000$) rad.

3

THE FUNDAMENTAL ELECTRICAL AND MAGNETIC UNITS

When units were first required for the measurement of electrical quantities it was natural to define them in terms of the three fundamental units, centimeter, gram, and second, which were already commonly used in mechanics. Electrical phenomena are related to mechanical phenomena by two effects: (*a*) the force between static electric charges (Coulomb's law) and (*b*) the force between electric currents (Ampere's law). Correspondingly, two distinct systems of cgs electrical units were introduced: the electrostatic and electromagnetic systems.

Neither of these systems has units of convenient size in practical applications. Consequently, a practical set of electrical units, defined as decimal multiples of the electromagnetic units was established by various International Congresses of Electricians meeting between 1881 and 1889. The first two units defined were the ohm (10^9 emu), chosen to be similar to the Siemens unit of resistance, and the volt (10^8 emu), chosen to be similar to the emf of the Daniell cell. From these, six other units, the ampere, coulomb, joule, watt, henry, and farad were derived. These practical units were not made into a complete system, since no magnetic units were defined, the unmodified magnetic units of the electromagnetic system (e.g., oersted and gauss) being used in practical applications.

In 1901, Giorgi showed that if the meter, kilogram, and second were used as fundamental units instead of the centimeter, gram and second, a single, consistent and comprehensive system of electrical and magnetic units could be built up incorporating the already firmly-established practical units. This is because, using the meter, kilogram and second, the unit of mechanical power becomes 10^7 erg s^{-1}, which is the appropriate practical electrical unit, that is, the watt. The use of the Giorgi system, also known as the mks system, or the Absolute Practical System was approved by an International Electro-technical Commission in 1935. The Absolute Practical System, with the ampere as the electrical base unit was adopted by the CGPM for SI.

4

RELATIONS BETWEEN THE SYSTEMS OF ELECTRICAL UNITS

Coulomb's law for the force F between charges Q_1 and Q_2, distance r apart in vacuo, may be expressed in the form

$$F = \frac{Q_1 Q_2}{\varepsilon_1 r^2} \qquad \qquad ...(1.1)$$

where ε_1 is a constant, called the permittivity of free space. In the cgs electrostatic system, ε_1 is chosen to be unity. This choice of the value of ε_1, together with the use of the centimeter, gram, and second uniquely determines the system of units.

Ampere's law for the force between two parallel current elements $I_1 ds_1$ and $I_2 ds_2$ distance r apart in vacuo may be expressed in the form

$$F = \mu_1 \frac{I_1 ds_1 I_2 ds_2 \sin\theta}{r^2} \qquad \qquad ...(1.2)$$

where μ_1 is a constant, called the permeability of free space. In the cgs electromagnetic system, μ_1 is chosen to be unity. This choice of the value of μ_1, together with the use of the centimeter, gram, and second, again determine the system of units uniquely.

It may be noted that these two systems of units, defined by $\varepsilon_1 = 1$ and $\mu_1 = 1$, cannot be combined directly to form a single consistent system. It can be shown from Maxwell's electromagnetic theory that, in any consistent system of units, $\mu_1 \varepsilon_1 = 1/c^2$, where c is the velocity of electromagnetic radiation (e.g., light) in free space, measured in the appropriate units of length and time (e.g., $c \simeq 3 \times 10^8$ ms^{-1}).

In SI, neither ε_1 nor μ_1 is chosen to be unity. The fundamental units chosen are the meter, kilogram, and second and ampere which are sufficient to determine the complete system uniquely. In particular, μ_1 may be shown to have the value 10^{-7} newton ampere^{-2}, where the newton is the SI unit of force. This value of μ_1 is readily derived from equation (2). The appropriate value of ε_1 is then calculated, knowing the experimentally determined value of the velocity of light.

5

RATIONALIZATION OF MKS UNITS

It is found that many formulae are simplified if the permeability of free space is re-defined as $\mu_0 = 4\pi\mu_1$. Ampere's law for current elements in free space is then expressed in "Rationalized mks units" as

$$F = \frac{\mu_0 I_1 ds_1 I_2 ds_2 \sin\theta}{4\pi\, r^2} \qquad \ldots(1.3)$$

where $\mu_0 = 4\pi \times 10^{-7}$ newton ampere^{-2} (or henry meter^{-1}).

Similarly, the permittivity of free space is re-defined as $\varepsilon_0 = \varepsilon_1/4\pi$, and Coulomb's law, for charges in free space, is expressed in rationalized mks units as

$$F = \frac{Q_1 Q_2}{4\pi\varepsilon_0 r^2} \qquad \ldots(1.4)$$

The value of ε_0, given by $1/c^2\mu_0$, is approximately 8.85×10^{-12} farad meter^{-1}.

For an isotropic, homogeneous medium other than free space, μ_0 in equation (3) is replaced by $\mu_0 = \mu_r\mu_0$, where μ_r, is the relative permeability of the medium; and ε_0 in equation (4) is replaced by $\varepsilon = \varepsilon_r \varepsilon_0$, where ε_r, is the relative permittivity (dielectric coefficient) of the medium.

6

DEFINITIONS OF ELECTRIC AND MAGNETIC QUANTITIES IN SI

The base unit

Current (*I*): The unit of current is the *ampere* (A) defined as that constant current which, if maintained in each of two infinitely long straight parallel wires of negligible cross-section placed 1 meter apart, in a vacuum, will produce between the wires a force of 2×10^{-7} newtons per meter length.

Derived units

Charge (*Q*): The unit of charge (quantity) is the *coulomb* (*C*) defined as the quantity of electricity transported per second by a current of 1 ampere.

Potential Difference (*V*): The unit of potential difference is the *volt* (V), defined as the difference of electrical potential between two points of a wire carrying a constant current of 1 ampere when the power dissipation between those points is 1 watt.

Resistance (*R*): The unit of resistance is the *ohm* (Ω), defined as the electrical resistance between two points of a conductor when a constant potential difference of 1 volt applied between these points produces in the conductor a current or 1 ampere.

Conductance (G): The unit of conductance is the *siemens* (*S*), defined as the electrical conductance between two points of a conductor when a constant potential difference of 1 volt applied between these points produces in the conductor a current of 1 ampere.

Inductance (L): The unit of inductance is the *henry* (*H*), defined as the inductance of a closed circuit in which an electromotive force of 1 volt is produced when the current in the circuit varies uniformly at the rate of 1 ampere per second.

Capacitance (C): The unit of capacitance is the *farad* (*F*), defined as the capacitance of a capacitor between the plates of which there appears a difference of potential of 1 volt when it is charged by 1 coulomb.

Magnetic Intensity (H): is defined through Ampere's theorem for the intensity due to a current element. In the usual notation $H = \dfrac{I.ds.\sin\theta.}{4\pi r^2}$ Unit *ampere meter^{-1}*.

Magnetic Flux (ϕ) of the induction B: is defined as $\int$ B.n dA where n is a unit vector perpendicular to an element of area dA. Unit, *weber* (Wb).

Magnetic Flux Density or Induction (B): is defined through the equation for the force on a current element placed in a magnetic field, viz $F = B.I.ds.\sin\theta$, in the usual notation. Unit, Tesla (T). $B = \mu_0 \mu_r H$ where μ_r is the relative permeability of the medium with respect to free space and μ_0 is the permeability of free space. $\mu_0 = 4\pi \times 10^{-7}$ henry meter^{-1}.

Magnetic Moment (m)*: is the couple exerted on a magnet placed at right angles to a uniform field with unit flux density. Unit, *ampere meter^2*.

Intensity of Magnetisation (M)*: is the magnetic moment per unit volume of a magnet. Unit, *ampere meter^{-1}*.

Pole Strength (P)*: On the mks system the hypothetical concept of an isolated magnetic pole is abandoned by many writers on the grounds that all magnetism arises from electrical effects, hence the definitions of *H* and *B* (above). Other writers use the idea of a magnetic pole as a simple and convenient concept in magnetometry. In this connection, we define a unit magnetic, pole as one which when situated 1 meter distant in vacuum from an equal pole experiences a force of $\mu_0/4\pi$ newtons. Alternatively, it can be defined as that pole strength which when placed in a unit induction experiences a force of 1 newton. Unit, *ampere-meter*.

<table>
<tr><td>NOTE</td><td>*For these definitions, we adopt the Sommerfeld system of units in which the magnetic moment of a current loop is the product of the area of the loop and the current flowing around the edge of the loop: m = IA. An alternative system due to Kennelly uses the relation $m = \mu_0 IA$.*</td></tr>
</table>

Line of Force: A line of force is a curve in a magnetic field, such that the tangent at every point is the direction of the magnetic force at that point.

Magnetomotive Force (F_m): is defined as the line integral $\int H.dl$ evaluated for a closed path. It is equal to the total conduction current linked. Unit, *ampere*.

Coulomb's Magnetic Law: states that the force between two poles P_1 and P_2 situated distance d apart is given by $F = \dfrac{\mu_r \mu_0 P_1 P_2}{4\pi d^2}$, where μ_r is the permeability of the medium and μ_0 the permeability of the free space = $4\pi \times 10^{-7}$ henry meter^{-1}.

$\mu_r\mu_0$ replaces the permeability μ of the cgs system.

Electrical Intensity (X or E): The electrical intensity at a point in an electric field is the force exerted on unit charge (1 coulomb) placed at that point, assuming that the field is not disturbed by so doing. Unit, *volt meter*$^{-1}$ (which is equivalent to the newton coulomb^{-1}).

Coulomb's Electrostatic Law: appears in the form $F = \dfrac{Q_1 Q_2}{4\pi \varepsilon_0 \varepsilon_r d^2}$, where Q_1 and Q_2 are the two charges situated a distance d apart in a medium whose permittivity relative to that of free space is ε_r. The permittivity of free space $\varepsilon_0 \simeq (1/36\pi) \times 10^{-9}$ farad meter^{-1}. ε_r is a pure number. (The product $\varepsilon_0 \varepsilon_r$ is analogous to the dielectric constant as defined in the cgs system.)

7

DEFINITIONS IN THE CGS ELECTROMAGNETIC SYSTEM OF UNITS

Magnetic Pole (*P*): When two like magnetic poles are placed 1 cm apart in vacuo, they repel one another with a force of 1 dyne.

Magnetic Field Strength or Intensity (*H*): is the force experienced by a unit North pole when placed at the given point in a magnetic field, it is assumed that the introduction of the pole does not disturb the field. Unit, *oersted*. The intensity is one oersted when a unit North pole experiences a force of 1 dyne on being placed at the given point in the field. The field strength in vacuum is represented as the number of lines of force passing perpendicularly through 1 cm² placed at the point in question. On this convention 4π lines of force leave a unit North pole.

Magnetic Flux (ϕ): through any area at right angles to a magnetic field is the product of the area and the field strength. Unit, *maxwell*. One maxwell is the flux through unit area (1 cm²) placed perpendicularly to a unit uniform field. Hence one line of force is equivalent to one maxwell.

Magnetic Moment (*m*): of a magnet, is the couple exerted on the magnet when placed at right angles to a unit uniform field. For a bar magnet, it is equivalent to the product of the pole strength and the distance between the poles. Unit, *pole cm*.

Magnetic Potential (Ω): is the work done in bringing a unit North pole from infinity or a point of zero potential to the point in question. Unit, *gilbert*. 1 gilbert is that potential that requires the expenditure of 1 erg of work in bringing a unit North pole from infinity to the point.

Intensity of Magnetisation (M): of a sample of material is the magnetic moment per unit volume.

Magnetic Susceptibility (χ): of a material is the ratio of the intensity of magnetization produced in the sample to the magnetic field which produced the magnetization, $\chi = \dfrac{M}{H}$.

N.B.: χ is not a constant but is a function of H.

Magnetic Induction (B): in any material is the number of lines of magnetic force (often called lines of induction) passing perpendicularly through the unit area. Unit, *gauss*. One gamma = 10^{-5} gauss.

Magnetic Permeability (μ): of any material is the ratio of the magnetic induction in the sample to the magnetic field producing it, that is, $\mu = B/H$. Although μ is so defined, B is not proportional to H, for $B = H + 4\pi M$. Also $B/H = 1 + 4\pi M/H$ or $\mu = 1 + 4\pi\chi$. Hence μ is not a constant but a function of H. (see χ above).

Coulomb's Law of Force: states that the force F between two poles of strength P_1 and P_2 is proportional to the product of the pole strengths and inversely proportional to the square of their distance apart (d). Thus $F = \dfrac{P_1 P_2}{\mu d^2}$ where 1/μ is the constant of proportionality, μ is the permeability of the medium in which the poles are located. In this system, as already stated, the permeability of free space is defined to be unity.

Current (I): The electromagnetic unit (emu) of current is that which when flowing around 1 cm arc of a circle of radius 1 cm, produces a magnetic field of 1 oersted at the center. Unit, *emu of current*.

Charge (Q): The emu of charge (quantity) is that delivered in 1 second by the passage of unit current. Unit, *biot*.

Potential Difference (PD): When unit current flows between two points in a circuit and unit work (1 erg) is done per second, the PD between the two points is unity. Unit, *emu of PD*.

Electromotive Force (emf): When lines of magnetic force cut a conductor an emf is created which is given numerically (in emu) by the number of lines cut per second. Emf = *dn/dt*.

Resistance (R): A conductor has unit resistance when applying unit PD, unit current flows. Unit, *emu of resistance*.

Self Inductance (L): A conductor possesses unit self-inductance if unit emf is developed across it when the rate of change of current is unity. Unit, *emu of self-inductance*.

Mutual Inductance (M): Two conductors possess unit mutual inductance when unit emf is developed in one by unit rate of change of current in the other. Unit, *emu of mutual inductance*.

Capacitance (C): A conductor has unit capacitance when the addition of unit charge raises its potential by unity. Unit, *cm*.

Dielectric Constant or Specific Inductivity Capacity (ε_r): of a material is the ratio of the capacity of a condenser with the material as dielectric to that of the same condenser in vacuum without a material dielectric.

Coulomb's Electrostatic Law of Force: states that the force F between two charges Q_1 and Q_2 is proportional to the product of the charges and inversely proportional to the square of their distance apart d. Thus $F = \dfrac{Q_1 Q_2}{\varepsilon_r d^2}$, where $1/\varepsilon_r$ is the constant of proportionality. ε_r is the dielectric constant of the medium in which the charges are located. On the electrostatic system of units, the dielectric constant of free space is unity.

8

HEAT UNITS AND DEFINITIONS

Temperature (t, θ or T). In SI, temperatures are measured on the thermodynamic scale with the Absolute Zero as zero and the triple point of water (*i.e.*, the temperature at which ice, water, and water vapor are in equilibrium) as the upper fixed point. The thermodynamic scale is that given by a theoretical Carnot heat engine and is equal to the perfect gas scale.

SI base unit, the Kelvin (K). The kelvin (K) unit of thermodynamic temperature, is the fraction 1/273·16 of the thermodynamic temperature of the triple point of water.

The Degree Celsius (°C). The centigrade scale of temperature used the ice point as zero and the boiling point of water at 1 standard atmosphere as the upper fixed point set to be 100°C. The Celsius scale of temperature is defined to be the same as the thermodynamic scale with the zero shifted to the ice point, which is 273.15 K, and thus:

$$\theta \,/\,°C = T\,/\,K - 273.15$$

The International Practical Scale of Temperature (IPST). In view of the difficulty of measuring on the thermodynamic scale, a scale of temperature based on fixed points was suggested by the 7th CCPM in 1927. The scale has been revised since so as to make temperatures on this scale agree as nearly as possible with the thermodynamic Celsius scale. A list of the fixed points and other important temperatures can be found in the book.

The Degree Fahrenheit (°F). On the Fahrenheit scale, the ice point is 32°F and the steam point, 212°F. Thus $t/°F = 32 + 1.8 (\theta/°C)$.

The Degree Reaumur (°R). On the Reaumur scale, the ice point is 0°R and the steam point, 80°R. Thus $t/°R = 0.8 (\theta/°C)$.

Quantity of Heat (Q). Quantities of heat are measured in joules (J) in SI. Other units have been used, notably the *calorie*. The calorie is the amount of heat required to raise the temperature of 1 gram of water by 1°C. This definition is not very precise however as the specific heat capacity of water varies with temperature. The 15°*calorie* was defined as the heat required to raise the temperature of 1 g of water from 14.5°C to 15.5°C. The *mean calorie* was defined as one-hundredth of the heat required to raise the temperature of 1 g of water from 0°C to 100°C. The *kilocalorie* (1 000 calories) has also been used and written Calorie. Where quantities of heat are expressed in calories, it is recommended that the conversion factor to convert to joules be stated.

In the fps system, the *British thermal unit* (Btu) is used. This is the quantity of heat required to raise the temperature of 1 lb of water through 1°F. The therm is 10^5 Btu.

Specific Heat Capacity (c_p, c_0). This is the amount of heat required to raise the temperature of 1 kg of a substance 1 K. Unit, $J\ kg^{-1}\ K^{-1}$.

Molar Heat Capacity (C_m). This is the amount of heat required to raise the temperature of 1 mol of substance through 1K. Unit, $Jmol^{-1}\ K^{-1}$.

Heat Capacity (C). The amount of heat required to raise the temperature of a body through 1 K. Units, $J\ K^{-1}$.

Water Equivalent. The mass of water having the same total heat capacity as the given body.

Thermal Conductivity (λ). The rate of flow of heat (dQ/dt) through a surface of area, A, in a medium is given by:

$$\frac{dQ}{dt} = -\lambda A \frac{dT}{dx},$$

where (dT/dx) is the temperature gradient, measured in the direction normal to the surface. The quantity λ is the thermal conductivity of the medium. Units, $Jm^{-1}s^{-1}K^{-1}$, or $Wm^{-1}K^{-1}$.

Specific Latent Heat (l). The specific latent heat of fusion (specific enthalpy Change on fusion) of a body is the heat required to convert 1 kg of the solid at its melting point into liquid at the same temperature. Unit, $J\ kg^{-1}$.

The specific latent heat of vaporization (enthalpy change on vaporization) of a liquid is the heat required to convert 1 kg of the liquid at its boiling point into vapor at the same temperature. Unit, $J\ kg^{-1}$.

Linear Expansivity (α). The increase in length per unit rise in temperature. Unit, K^{-1}.

Cubic Expansivity (γ). The increase in volume per unit rise in temperature. Unit, K^{-1}.

Critical Temperature (T_c) of a gas or vapor is that temperature above which it is not possible to liquefy the gas by the application of pressure alone. To liquefy a gas it must be cooled below its critical temperature before being compressed.

Critical Pressure (p_c): That pressure which just liquefies a gas at its critical temperature.

Critical Volume (V_c): The volume of unit mass of gas at its critical temperature and pressure, that is, it is the reciprocal of the critical density. It is often taken as the volume of one mole of a gas at its critical temperature and pressure.

Radiation. Stefan-Boltzmann Law: The total energy, E, of all wavelengths radiated per second per square meter by a full radiator at temperature T to surroundings at T_0 is given by $E = \sigma(T^4 - T_0^4)$, where σ is Stefan's constant, $\sigma = 5.669\ 7 \times 10^{-8}\ W\ m^{-2}\ K^{-4}$.

Planck's Radiation Law: The energy density of radiation in an enclosure at temperature T having wavelengths in the range λ to $\lambda + d\lambda$ is $u\lambda\ d\lambda$, where

$$u_\lambda\ d_\lambda = 8\pi ch\lambda^{-5}\ (\exp hc/\lambda kT - 1)^{-1}\ d\lambda = c_1\lambda^{-5}(\exp c_2/\lambda T - 1)^{-1}\ d\lambda$$

$$c_1 = 4.992\ 1 \times 10^{-24}\ Jm \qquad c_2 = 1.43879 \times 10^{-2}\ m\ K$$

The corresponding relation for radiation of frequency, v, is

$$u_v \, d_v = (8\pi h/c^3) \, (\exp hv/kT - 1)^{-1} \, v^3 \, dv.$$

h = Planck's constant;

c = speed of light;

k = Boltzmann's constant;

T = temperature of the enclosure.

Wien's Displacement Law: The wavelength of the most strongly emitted radiation in the continuous spectrum from a full radiator is inversely proportional to the absolute temperature of that body, that is, $\lambda T = b$, where b is Wien's constant = 2.898×10^{-3} m K.

The Energy (E) of a quantum of radiation of frequency v is $E = hv$ where h is Planck's constant.

9

PHOTOMETRIC AND OPTICAL UNITS AND DEFINITIONS

Luminous intensity. In SI, the unit of luminous intensity is the candela. This unit replaces the International candle which was defined in terms of the light emitted per second in all directions by a specified electric lamp.

SI base unit, the candela (cd). The candela is the luminous intensity, in the perpendicular direction, of a surface of 1/600 000 square meters of a full radiator at the temperature of freezing platinum under a pressure of 101 325 newtons per square meter.

1 candela = 0.982 international candles.

Luminous flux: The unit of luminous flux, the lumen (1m) is defined as the light energy emitted per second within, unit solid angle by a uniform point source of unit luminous intensity. Thus, 1 cd = 1 1m sr^{-1}.

Illuminance of a surface is defined as the luminous flux reaching it perpendicularly per unit area. The British unit is the lumen ft^{-2}, formerly called the foot-candle (fc). The metric unit is the lumen m^{-2} or lux (1x).

Lambert's Cosine law: For a surface receiving light obliquely, the illumination is proportional to the cosine of the angle which the light makes with the normal to the surface.

Brightness of a surface is that property by which the surface appears to emit more or less light in the direction of view. This is a subjective quantity. The corresponding physical measurement of the light actually emitted is called luminance.

Luminance of a surface is the measure of the light actually emitted (i.e., the luminous intensity) per unit projected area of surface, the plane of projection being perpendicular to the direction of view. Unit, cd ft^{-2} or cd m^{-2}. In engineering, the luminance of an ideally diffusing surface emitting or reflecting one-lumen ft^{-2} is called one foot-lambert (ft–L).

The Refractive index of a material (n) is the ratio of the velocity of light in free space to that in the material.

Snell's law: For light incident on a boundary between two media, the ratio of the sine of the angle of incidence (the angle between the light ray in the first medium and the normal to the boundary surface) to the sine of the angle of refraction (the angle between the refracted ray in the second medium and the normal) is a constant, being equal to the inverse ratio of the refractive indices of the two media.

Dioptre is the unit of measure of the power of a lens and is given numerically by the reciprocal of the focal length expressed in meters.

CHAPTER 10

ACOUSTICAL UNITS AND DEFINITIONS

Pressure: The unit of sound pressure is the pascal usually quoted as the root mean square (r.m.s.) pressure for a pure sinusoidal wave.

Frequency: The unit of frequency is the cycle per second, now designated the hertz (Hz).

Threshold of Hearing is, for a normal (average) observer, the sound level or intensity which is just audible. For a pure sinusoidal note of frequency 1000 Hz, it is close to a root mean square pressure of 2×10^{-5} Pa.

Power Ratio: The unit of acoustical (or electrical) power measurement with respect to a standard level, is one *bel*. The interval between two powers W_1 and W_0 in bels is $\log_{10}(W_1/W_0)$. In practical work, the decibel (dB) is used. The interval between two powers W_1 and W_0 is $10\log_{10}(W_1/W_0)$ dB. In some instances, it is more convenient to employ natural logarithms. The power ratio so obtained is called the neper and is defined as follows. The power interval between W_1 and W_0 is $\frac{1}{2}\log_e(W_1/W_0)$ nepers. Hence 1 neper = 8.686 dB.

Intensity (I) of a sound wave in a given direction is the sound energy transmitted per second in this direction through unit area placed perpendicularly to the specified direction. Unit, $\mathrm{W\ m^{-2}}$. For a sinusoidal plane or spherical wave, the intensity is proportional to the mean square pressure exerted on an area at right angles to the given direction. Hence the interval between two intensities is given by $10 \log_{10} (I_1 / I_0)$ dB or $20 \log_{10} (\rho_1/\rho_0)$ dB where ρ_1 and ρ_0 are the r.m.s. pressures corresponding to the intensities I_1 and I_0.

Loudness is the physiological counterpart of acoustical intensity. It is a function of the intensity but also varies with frequency and composition of the note being heard. The *Weber–Fechner Law* states that the sensation (loudness) is proportional to the logarithm of the stimulus (intensity).

Loudness level of a sound is judged by comparison in free air with a standard sinusoidal note whose frequency is 1000 Hz. The unit is the *phon*. If an average observer decides that a sound is equally loud as the standard 1000 Hz note of intensity n dB above the standard reference level corresponding to a r.m.s. pressure of 2×10^{-5} Pa (i.e., the threshold of hearing), then the sound is said to have an "equivalent loudness" of n British Standard phons.

Reverberation in an enclosure is the persistence of sound due to multiple reflections from the walls, etc. of the enclosure.

Reverberation time is the time required, from the moment of cessation of a sound for the intensity to, drop by 60 dB, that is, to one-millionth of its original value. Unit, *second*.

Absorption Coefficient of a surface is the ratio of the sound energy absorbed to the total sound energy incident on the surface. The ideal absorber is one from which no sound is reflected or scattered. For unit area of various substances, the coefficient is expressed in terms of equivalent area of open window (diffraction effects excluded). Unit, *ft² of open window or Sabine*. The coefficient varies with frequency.

Sabine's Relation: For an auditorium whose walls, etc. consist of areas S_1 S_2, etc. of absorption coefficient α_1 α_2, etc., the reverberation time t (in seconds) is given by $t = \dfrac{0.05V}{\Sigma a\ S}$ (unit of length, ft) or $t = \dfrac{0.16V}{\Sigma a\ S}$ (unit of length, meter) where V is the volume of the auditorium and $\Sigma \alpha\ S = \alpha_1 S_1 + \alpha_2 S_2 +$ etc.

11

PROPERTIES OF THE ELEMENTS

The following table lists the elements with atomic number up to 92 alphabetically by name. Columns 1-4 and 13-16 are self-explanatory. Column 5 gives the crystal structures of the elements in their solid state. Where a change in structure occurs, the transition temperature is indicated (in K) under the crystal structures. The following abbreviations are used:

bcc = body-centered cubic   hex = hexagonal

cubic (diam) = diamond structure   mon = monoclinic

fcc = face-centered cubic   ortho = orthorhombic

hcp = hexagonal close-packed   tetra = tetragonal

Column 6 lists the atomic radii of the elements in pm (10^{-12} m). These radii are calculated as half the distance of closest approach of atomic centers in the crystalline state. Column 7 gives the principal oxidation numbers and column 8, the corresponding ionic radii. Columns 9 and 10 give the energies (eV) required to remove the first and second electrons from the atom multiply by 96.49 to convert to kJ mol^{-1}. Column 11 gives the energy required to remove an electron from the negative ion formed by the atom with an extra electron. These "electron affinities" are difficult to measure and there are few reliable results. Column 12 gives the electronegativities assigned to the elements by Pauling. These are numbers between 0 and 4 which may be used in determining the contribution of the ionic and covalent components of the bonds between different atoms.

Symbols	Name	Atomic number, Z	Atomic weight (M/g mol⁻¹)	Electron affinities (E_i/eV)	Electronegativities	Density (ρ/kg m⁻³)	Melting point (T_m/K)	Boiling point (T_b/K)
Al	Aluminum	13	26.98	0.5	1.5	2 700	933.2	2 740
Sb	Antimony	51	121.75	> 2.0	1.9	6 700	903.7	1 650
Ar	Argon	18	39.95	-1.0	—	1.66	83.7	87.4
As	Arseni	33	74.92	—	2.0	5 730	1 090 (28 atm)	886 (sub)
Ba	Barium	56	137.34	—	0.9	3 600	1 000	1 910
Be	Beryllium	4	9.01	0.30	1.5	1 800	1550	3 243
Bi	Bismuth	83	208.98	> 0.7	1.9	9 800	544.4	1 830
B	Boron	5	10.81	0.33	2.0	2 500	2 600	2 820
Br	Bromine	35	79.90	3.363	2.8	3 100 (298 K)	265.9	331.9 (sub)
Cd	Cadmium	48	112.40	—	1.7	8 650	594.2	1 038
Cs	Caesium	55	132.90	> 0.19	0.7	1 870	301.6	960
Ca	Calcium	20	40.08	—	1.0	1 540	1 120	1 760
C	Carbon	6	12.01	1.25	2.5	2 300	>3 800	5 100
Cl	Chlorine	17	35.45	3.615	3.0	3.21 (273 K)	172.1	238.5
Cr	Chromium	24	52.00	0.98	1.6	7 200	2 160	2 755
Co	Cobalt	27	58.93	0.9	1.8	8 900	1 765	3170
Cu	Copper	29	63.55	1.8	1.9	8 930	1 356	2 868

F	Fluorine	9	19.00	3.448	4.0	1.7 (273 K)	53.5	85.01
Ge	Germanium	32	72.59	—	1.8	5 400	1 210.5	3 100
Au	Gold	79	196.97	2.1	2.4	19 300	1 336.1	3 239
Hf	Hafnium	72	178.49	—	1.3	13 300	2 423	5 700
He	Helium	2	4.003	−0.53	—	0.166	0.95 (26 atm)	4.21
H	Hydrogen	1	1.00797	0.76	2.1	0.08987 (273 K)	14.01	20.4
In	Indium	49	114.82	—	1.7	7 310	429.8	2 300
I	Iodine	53	126.90	3.070	2.5	4 940	386.6	457.4
Ir	Iridium	77	192.2	—	2.2	22 420	2 716	4 800
Fe	Iron	26	55.85	0.6	1.8	7 870	1 808	3 300
Kr	Krypton	36	83.80	—	—	3.49	116.5	120.8
La	Lanthanum	57	138.91	—	1.1	6 150	1190	3 742
Pb	Lead	82	207.19	—	1.8	11 340	600.4	2 017
Li	Lithium	3	6.94	0.82	1.0	534	452	1 590
Mg	Magnesium	12	24.31	−0.32	1.2	1 741	924	1 380
Mn	Manganese	25	54.94	—	1.5	7 440	1 517	2 370
Hg	Mercury	80	200.59	1.54	1.9	13 590	234.3	629.7
Mo	Molybdenum	42	95.94	1.0	1.8	10 200 (273 K)	2 880	5 830

Symbols	Name	Atomic number, Z	Atomic weight (M/g mol^{-1})	Electron affinities (E_a/eV)	Electronegativities	Density (P/kgm^{-3})	Melting point (T_m/K)	Boiling point (T_b/K)
Ne	Neon	10	20.18	– 0.57	—	0.839	24.5	27.2
Ni	Nickel	28	58.71	1.3	1.8	8 900	1 726	3 005
Nb	Niobium	41	92.91	—	1.6	8.570	2 741	5 200
N	Nitrogen	7	14.01	0.05	3.0	1.165	63.3	77.3
O	Oxygen	8	16.00	1.471	3.5	1.33	54.7	90.2
Pd	Palladium	46	106.4	—	2.2	12 000	1 825	3 200
P	Phosphorus	15	30.97	0.8	2.1	2 200 (r) 1 800 (y)	317.2	552
Pt	Platinum	78	195.09	—	2.2	21 450	2 042	4 100
Po	Polonium	84	209	—	2.0	9 400	527	1 235
K	Potassium	19	39.10	0.82	0.8	860	336.8	1 047
Ra	Radium	88	226	—	0.9	5 000	970	1 410
Rn	Radon	86	222	—	—	9.73 (273 K)	202	211.3
Re	Rhenium	75	186.2	0.15	1.9	20 500	3 450	5 900
Rh	Rhodium	45	102.91	—	2.2	12 440	2 230	4 000
Sm	Samarium	62	150.35	—	1.1	7 500	1 345	2 200
Se	Selenium	34	78.96	3.7	2.4	4 810	490	958
Si	Silicon	14	28.09	1.5	1.8	2 300	1 680	2 628
Ag	Silver	47	107.87	2.5	1.9	10 500	1 234	2 485

Na	Sodium	11	22.99	0.84	0.9	970	371	1 165
Sr	Strontium	38	87.62	—	1.0	2 600	1 042	1 657
S	Sulfur	16	32.06	2.07	2.5	2 070	386	717.7
Ta	Tantalum	73	180.95	—	1.5	16 600	3 269	5 698
Te	Tellurium	52	127.60	3.6	2.1	6 240	722.6	1 260
Tl	Thallium	81	204.37	—	1.8	11 860	576.6	1 730
Th	Thorium	90	232.04	—	1.3	11 500	2 000	4 500
Sn	Tin	50	118.69	—	1.8	7 300	505.1	2 540
Ti	Titanium	22	47.90	0.39	1.5	4 540	1 948	3 530
W	Tungsten	74	183.85	0.5	1.7	19 320	3 650	6 200
U	Uranium	92	238.03	0.94	1.7	19 050	1 405.4	4 091
V	Vanadium	23	50.94	—	1.6	6 100	2 160	3 300
Xe	Xenon	54	131.30	—	—	5.50	161.2	166.0
Y	Yttrium	39	88.91	—	1.2	4 600	1 768	3 200
Zn	Zinc	30	65.37	—	1.6	7 140	692.6	1 180
Zr	Zirconium	40	91.22	—	1.4	6 500	2 125	3 851

CHAPTER 12

PROPERTIES OF METALLIC SOLIDS (AT 293 K)

Values quoted for Tensile Strength and Yield Stress are in units of 10^4 N m^{-1} (= MPa). Values of Young's Modulus are in units of 10^9 N m^{-1} (= GPa). These values are typical observations and are approximate only. The elastic properties vary somewhat between specimens depending on the manufacturing process and the previous history of the sample. The Shear Modulus (G) and Bulk Modulus (K) can be calculated from the relations: $G = \frac{1}{2}E/(1 + v)$ and $K - \frac{1}{2}E/(1 - 2v)$, where E is Young's Modulus and v is Poisson's Ratio.

	Name	Density ρ/kg m^{-3}	Melting point T_m/K	Specific latent heat of fusion l/J kg^{-1} ×10^4	Specific heat capacity C_p/J kg^{-1} K^{-1}	Linear expansivity α/K^{-1} ×10^{-6}	Thermal conductivity λ/W m^{-1} K^{-1}	Electrical resistivity ρ/Ω m ×10^{-8}	Temperature coefficient of resistance $(1/\rho_0)$ $(d\rho/dT)$ /K^{-1} ×10^{-4}	Tensile strength $(\sigma_T$/MPa)	Yield strength $(\sigma_y$/MPa)	Elongation (e/%)	Young's Modulus (E/GPa)	Poisson's ratio	
1.	Aluminum	2 710	932	38	913	23	201	2.65	40	80	50	43	71	0.34	1
2.	Aluminum strong alloy	2 800	800	39	880	23	180	5	16	600	550	10			2
3.	Antimony	6 680	904	16	205	10	18	40	~50				78		3
4.	Bismuth	9 800	544	5	126	13	8	115	45				32	0.33	4
5.	Brass (70Cu/30Zn)	8 500	1300		370	18	110	~8	~15	550	450	8	100	0.35	5
6.	Bronze (90Cu/10Sn)	8 800	1300		360	17	180	30		260	140	10			6
7.	Cobalt	8 900	1765	25	420	12	69	6	66	~500					7
8.	Constantan	8 880	1360		420	17	23	47	±0.4				170	0.33	8
9.	Copper	8 930	1356	21	385	17	385	1.7	39	150	75	45	117	0.35	9

No.															
10.	German silver (60Cu/25Zn/15Ni)	8 700	1300		400	18	29	33	4	450			130	0.33	10
11.	Gold	19 300	1340	7	132	14	296	2.4	34	120		40	71	0.44	11
12.	Invar (64Fe/36Ni)	8 000	1800	27	503	0.9	16	81	20	480	280	40	145	0.26	12
13.	Iron, pure	7 870	1810	10	106	12	80	10	65	300	165	45	206	0.29	13
14.	Iron, cast grey	7 150	1500	14	500	11	75	10		100			110	0.27	14
15.	Iron, cast white	7 700	1420	14		11	75	10		230		~ 0			15
16.	Iron, wrought	7 700	1810	14	480	12	60	14	60	~ 370	150	45	197	0.28	16
17.	Lead	11 340	600	2.6	126	29	35	21	43	15	12	50	18	0.44	17

	Name	Density ρ/kg m⁻³	Melting point T_f/K	Specific latent heat of fusion l/J kg⁻¹ ×10⁴	Specific heat capacity C_p/J kg⁻¹ K⁻¹	Linear expansivity α/K⁻¹ ×10⁻⁶	Thermal conductivity λ/W m⁻¹ K⁻¹	Electrical resistivity (ρ/Ω m) ×10⁻⁸	Temperature coefficient of resistance $(1/\rho_0)(d\rho/dT)$ K⁻¹ ×10⁻⁴	Tensile strength σ_t/MPa	Yield strength σ_y/MPa	Elongation e/%	Young's modulus E/GPa	Poisson's ratio
18.	Magnesium	1 740	924	38	246	25	150	4	43	190	95	5	44	0.29
19.	Manganin	8 500	1320	41	400	18	22	45	± 0.1	520	240		120	0.33
20.	Monel (70 Ni/30Cu)	8 800	1600			14	210	42	20	300	60	40		
21.	Nickel	8 900	1726	31	460	13	59	59	60			30	207	0.36
22.	Nickel, a strong alloy	8 500			380					1300	1200	10	110	0.38
23.	Phosphor bronze					17		7	60	560	420		120	0.38
24.	Platinum	21 450	2042	11	136	9	69	11	38	350			150	0.38
25.	Silver	10 500	1230	10	235	19	419	1.6	40	150	180	45	70	0.37
26.	Sodium	970	371	12	1240	71	134	4.5	44					

27.	Solder, soft (50Pb/50Sn)	9 000	490	190	176					45		50			27
28.	Stainless Steel (18Cr/8Ni)	7 930	1800		510	16	150	96	6	600	230	60			28
29.	Steel, mild	7 860	1700		420	15	63	15	50	460	300	35	210	0.29	29
30.	Steel, piano wire	7 800	1700				50			3000			210	0.29	30
31.	Tin	7 300	505	6.0	226	23	65	11	50	30			40	0.36	31
32.	Titanium	4 540	1950		523	9	23	53	38	620	480	20		0.36	32
33.	Zinc	7 140	693	10	385	31	111	5.9	40	150		50	110	0.25	33

13

PROPERTIES OF NON-METALLIC SOLIDS (AT 293 K)

The following table lists materials which do not readily conduct electricity. In many cases, the physical constants cannot be specified accurately as the values observed depend so much on the manufacture and life history of the specimen. The values given are to be taken as representative only.

	Name	Density ρ/kg m^{-3}	Melting point T_m/K	Specific heat capacity c_p/J kg^{-1} K^{-1}	Linear expansivity α/K^{-1} $\times 10^{-4}$	Thermal conductivity λ/W m^{-1} K^{-1}	Tensile strength σ_y/MPa	Elongation e/%	Young's modulus E/GPa	
1.	Alumina, ceramic	3 800	2 300	800	9	29	~150		345	1
2.	Bone	1 850					140		28	2
3.	Brick, building	2 300			9	0.6	~5			3
4.	Brick, fireclay	2 100			4.5	0.8				4
5.	Brick, paving	2 500			4.0					5

6.	Brick, silica	1 750				0.8			6
7.	Carbon, graphite	2 300	3 800	710	7.9	5.0			207	7
8.	diamond	3 300		525	~ 0	900			1 200	8
9.	Concrete	2 400		3 350	12	0.1	~ 4		14	9
10.	Cork	240		2 050		0.05				10
11.	Cotton	1 500		1 400			400			11
12.	Epoxy resin	1 120		1 400	39		50	2–6	4.5	12
13.	Fluon (PTFE)	2 200		1050	55	0.25	22	50–75	0.34	13
14.	Glass (crown)	2 600	1 400	670	9	1.0	~ 100		71	14
15.	Glass (flint)	4 200	1 500	500	8	0.8			80	15
16.	Glass (wool)	50	1 400	670		0.04				16
17.	Ice	920	273	2 100	51	2.0				17
18.	Kapok	50				0.03				18
19.	Magnesium oxide	3 600	3 200	960	12				207	19

20.	Marble	2 600		880	10	2.9				20
21.	Melamine formaldehyde	1 500		1 700	40	0.3	70		9	21
22.	Naphthalene	1 150	350	1 310	107	0.4				22
23.	Nylon	1 150	470	1 700	100	0.25	70	60–300		23
24.	Paraffin wax	900	330	2 900	110	0.25				24
25.	Perspex	1 190	350	1 500	85	0.2	50	2–7	3	25
26.	Phenol formaldehyde	1 300		1 700	40	0.2	50	0.4–0.8	6.9	26
27.	Polyethylene (low den)	920	410	2 300	250		13	400–800	0.18	27
28.	(high den)	955	410	2 300	250		26	100–300	0.43	28
29.	Polypropylene	900	450	2 100	62		35	>220	1.2	29
30.	Polystyrene	1 050	510	1 300	70	0.08	50	1–3	3.1	30
31.	Polyvinylchloride (non-rigid)	1 250	485	1 800	150		15	200–400	0.01	31
32.	Polyvinylchloride (rigid)	1 700	485	1 000	55		60	5–25	2.8	32

	Name	Density ρ/kg m^{-3}	Melting point (T_m/K)	Specific heat capacity c_p/J kg^{-1} K^{-1}	Linear expansivity α/K^{-1} $\times 10^{-4}$	Thermal conductivity λ/W m^{-1} K^{-1}	Tensile strength σy/MPa	Elongation e/%	Young's modulus E/GPa	
33.	Polyvinylidene chloride		470		190		30	160–240		33
34.	Quartz fiber	2 660	2 020	788	0.4	9.2			73	34
35.	Rubber (polyisoprene)	910	300	1 600	220	0.15	17	480–510	0.02	35

36.	Silicon carbide	3 170				4.5				36
37.	Sulfur	2 070	386	730	64	0.26				37
38.	Titanium carbide	4 500			7	28			345	38
39.	Wood, oak (with grain)	650				0.15			12	39
40.	Wood Spruce (with grain)	600							14	40
41.	Wood Spruce (across grain)								0.5	41

PROPERTIES OF LIQUIDS (AT 293 K)

Name	Density ρ/kg m^{-3}	Melting point T_m/K	Boiling point T_b/K	Specific latent heat of vaporization l/J kg^{-1} ×10^{-4}	Specific heat of capacity c_p/J kg^{-1} K^{-1}	Cubic expansivity γ/K^{-1} ×10^{-4}	Thermal conductivity λ/W m^{-1} K^{-1}	Surface tension σ/N m^{-1} ×10^{-3}	Viscosity η/N s m^{-2} ×10^{-3}	Refractive index n	Bulk modulus of rigidity K/GPa	
1. Acetic acid ($C_2H_4O_2$)	1049	290	391	39	1960	10.7	0.180	27.6	1.219	1.3718	2.49	1
2. Acetone (C_3H_6O)	780	178	330	52	2210	14.3	0.161	23.7	0.324	1.3620 (288 K)	~0.8	2
3. Benzene (C_6H_6)	879	279	353	40	1700	12.2	0.140	28.9	0.647	1.5011	1.10	3
4. Bromine (Br)	3100	266	332	18.3	460	11.3		41.5	0.993	1.66	1.58	4
5. Carbon disulfide (CS_2)	1293	162	319	36	1000	11.9	0.144	32.3	0.375	1.6276	1.16	5
6. Carbon tetrachloride (CCl_4)	1632	250	350	19	840	12.2	0.103	26.8	0.972	1.4607	1.12	6
7. Chloroform ($CHCl_3$)	1490	210	334	25	960	12.7	0.121	27.1	0.569	1.4467	1.1	7
8. Ether, diethyl ($C_4H_{10}O$)	714	157	308	35	2300	16.3	0.127	17	0.242	1.3538	0.69	8

No.	Liquid												No.
9.	Ethyl alcohol (C_2H_6O)	789	156	352	85	2500	10.8	0.177	22.3	1.197	1.3610	1.32	9
10.	Glycerol ($C_3H_8O_3$)	1262	293	563	83	2400	4.7	0.270	63	1495	1.4730	4.03	10
11.	Mercury (Hg)	13546	234	630	29	140	1.82	7.96	472	1.552	1.73	26.2	11
12.	Methyl alcohol (CH_4O)	791	179	337	112	2500	11.9	0.201	22.6	0.594	1.3276	0.97	12
13.	Nitrobenzene ($C_6H_5NO_2$)	1175	279	484	33	1400	8.6	0.160	43.9	2.03	1.5530	2.2	13
14.	Olive oil	920		570		1970	7.0	0.170	32	84	1.48	1.60	14
15.	Paraffin oil	800				2130	900	0.150	26	~1000	1.43	1.62	15
16.	Phenol (C_6H_6O)	1073	314	455	53	2350	7.9		40.9	12.74	1.5425 (313 K)		16
17.	Toluene (C_7H_8)	867	178	384	35	1670	10.7	0.134	28.4	0.585	1.4969	1.09	17
18.	Turpentine	870	263	429	29	1760	9.7	0.136	27	1.49	1.48	1.28	18
19.	Water (H_2O)	998	273	373	226	4190	2.1	0.591	72.7	1.000	1.333	2.05	19
20.	Water, sea	1025	264	~377		3900					1.343		20

PROPERTIES OF GASES AT STP

	Substance	Density ρ/kg m^{-3}	Boiling point T/K	Specific latent heat of vaporization l/J kg^{-1} ($\times 10^{-4}$)	Specific Heat Capacity c_p/J kg^{-1} K^{-1}	Ratio of specific heats $\gamma = c_p/c_v$	Thermal conductivity (λ/W m^{-1} K^{-1}) ($\times 10^{-4}$)	Viscosity η/N s m^{-2} ($\times 10^{-6}$)	Refractivity ($n-1$) ($\times 10^{-6}$)	Critical temperature T_c/K	Critical pressure (P_c/MPa)	Critical volume (V_c/m^3 mol^{-1}) ($\times 10^{-6}$)	
1.	Acetylene (C_2H_2)	1.173	189		1590	1.26	184	9.35	606	309	6.14	113*	1
2.	Air	1.293	83	21.4	993	1.402	241	18.325 (300 K)	292	132	3.77		2
3.	Ammonia (NH_3)	0.771	240	137.1	2190	1.310	218	9.18	376	405	11.3	72.5	3
4.	Argon (Ar)	1.784	87	15.8	524	1.667	162	21	281	151	4.86	75.2	4
5.	Carbon dioxide (CO_2)	1.977	195	36.4	834	1.304	145	14	451	304	7.38	94.0	5
6.	Carbon monoxide (CO)	1.250	81	21.1	1050	1.404	232	16.6	338	134	3.50	93.1	6
7.	Chlorine (Cl_2)	3.214	238	28.1	478	1.36	72	12.9	773	417	7.71	124	7
8.	Cyanogen (C_2N_2)	2.337	252	43.2	1720	1.26		9.28	835	401	6.0		8

9.	Ethylene (C_2H_4)	1.260	170	48.4	1500	1.26	164	9.7	696	283	5.12	127.4	9
10.	Helium (He)	0.179	4.25	2.5	5240	1.66	1415	18.6	36	5.3	0.23	58	10
11.	Hydrogen (H_2)	0.090	20.35	45.3	14300	1.41	1684	8.35	132	33.3	12.94	65.5	11
12.	Hydrogen chloride (HCl)	1.640	189	41.4	796	1.40		13.8	447	325	8.26	87	12
13.	Hydrogen sulfide (H_2S)	1.538	211	55.3	1020	1.32	120	11.7	634	374	9.01	97.9	13
14.	Methane (CH_4)	0.717	109	51.1	2200	1.313	302	10.3	444	191	4.62	98.7	14
15.	Nitric oxide (NO)	1.340	121	46.2	972	1.394	238	17.8	297	179	6.5		15
16.	Nitrogen (N_2)	1.250	77	20.9	1040	1.404	243	16.7	297	126	3.39	90.1	16
17.	Nitrous oxide (N_2O)	1.978	183	37.6	892	1.303	151	13.5	516	310	7.24	96.7	17
18.	Oxygen (O_2)	1.429	90	24.3	913	1.40	244	19.2	272	154	5.08	78	18
19.	Sulfur dioxide (SO_2)	2.927	263	40.3	645	1.26	77	11.7	686	430	7.88	122	19
20.	Water Vapor (273 K) (H_2O)	0.800		226.1	2020 (373 K)		158	8.7	254	647	22.12	56.8	20

*The critical volume is here defined as the volume of one mole of the gas at its critical temperature and pressure.

CHAPTER 16

MECHANICAL DATA

The Mohs Scale of Hardness

Substance	Hardness	Substance	Hardness	Substance	Hardness
Talc	1	Felspar	6	Fused zirconia	11
Gypsum	2	Vitreous silica	7	Fused alumina	12
Calcite	3	Quartz	8	Silicon carbide	13
Fluorite	4	Topaz	9	Boron carbide	14
Apatite	5	Garnet	10	Diamond	15

Approximate Hardness of Some Common Materials

Substance	Hardness	Substance	Hardness	Substance	Hardness
Agate	6–7	Calcium	1.5	Glass	4.5–6.5
Aluminum	2–3	Carborundum	9–10	Marble	3–4
Amber	2–2.5	Chromium	9	Penknife blade	6.5
Asbestos	5	Copper	2.5–3	Silver	2.5–2.7
Brass	3–4	Finger nail	2.5	Steel (mild)	4–5

Viscosities of Liquids and Their Temperature Dependence, η, N_s m^{-2}

Substance	0°C	10°C	20°C	30°C	40°C	50°C
Water	0.001787	0.001304	0.001002	0.00080	0.000653	0.000547
Aniline	0.0102	0.0065	0.0044	0.00316	0.00237	0.00185
Benzene	0.000912	0.0000758	0.000652	0.000564	0.000503	0.000442
Ethanol	0.00177	0.00147	0.0012	0.00100	0.000834	0.00070
Glycerol (propane 1, 2, 3-triol)	10.59	3.44	1.34	0.629	0.289	0.141
Rape oil	2.53	0.385	0.163	0.096	—	—

EMF of Standard Cells

Weston (Cadmium) cell (20°C)	=	1.0186 volts (absolute)
	=	1.0183 volts (international)
Clark cell (15°C)	=	1.4333 volts (absolute)
	=	1.4328 volts (international)

Temperature dependence

Weston cell

$$E_t = 1.0186 - 0.0000406\,(t - 20) - 9.5 \times 10^{-7}(t - 20)^2 \text{ absolute volts}$$

Clark cell

$$E_t = 1.4333 - 0.00119\,(t - 15) - 7 \times 10^{-6}(t - 15)^2 \text{ absolute volts}$$

Approximate EMFs of Cells

Bichromate	2 volts	Accumulator 2.0 volts (Ranges 1.85–2.2 volts)	
Bunsen	1.9 volts	Dry cell	1.5 volts
Daniell	1.08 volts	Nickel-Cadmium	1.3 volts
Grove	1.8 volts	Nickel-Iron	1.4 volts
Leclanché	1.46 volts	Zinc-Silver oxide	1.8 volts

Relative Permittivities (ε_r) of Various Substances at Room Temperature (293 K)

Solid	ε_r	Liquid	ε_r	Gas	ε_r
Amber	2.8	Acetone	21.3	Air	1.000536
Ebonite	2.7–2.9	Benzene	2.28	Argon	1.000545
Glass	5–10	Carbon tetrachloride	2.17	Carbon dioxide	1.000986
Ice (268 K)	75	Castor oil	4.5	Carbon monoxide	1.00070
Marble	8.5	Ether	4.34	Deuterium	1.000270
Mica	5.7–6.7	Ethyl alcohol	25.7	Helium	1.00007
Paraffin wax	2–2.3	Glycerine	43	Hydrogen	1.00027
Perspex	3.5	Medicinal paraffin	2.2	Neon	1.000127
Polystyrene	2.55	Nitrobenzene	35.7	Nitrogen	1.000580
P.V.C.	4.5	Pentane	1.83	Oxygen	1.00053
Shellac	3–3.7	Silicon oil	2.2	Sulfur dioxide	1.00082
Sulfur	3.6–4.3	Turpentine	2.23	Water vapor	1.00060
Teflon	2.1	Water	80.37	(393 K)	

Values given in the table above refer to low frequencies, gases at 1 atmosphere pressure.

Temperature—EMF Data for Thermocouples

The table gives the emf in millivolts for "hot junction" temperature from 0° to 100°C. The "cold junction" is maintained at 0°C.

Thermocouple	0°	10°	20°	30°	40°	50°	60°	70°	80°	90°	100°
Platinum–Platinum	0	0.06	0.11	0.17	0.23	0.30	0.36	0.43	0.50	0.57	0.64
(90%), Rhodium (10%)											
Copper–Constantan	0	0.39	0.79	1.19	1.61	2.03	2.47	2.91	3.36	3.81	4.28
Iron–Constantan	0	0.52	1.05	1.58	2.12	2.66	3.20	3.75	4.30	4.85	5.40

Magnetic Mass Susceptibility, χ_m (at 293 K)

The mass susceptibility is given by the expression, $\chi_m = (\mu_r - 1)/\rho$; where μ_r is the relative permeability and ρ the density of the specimen.

	χ_m/m^3		χ_m/m^3		χ_m/m^3
	$\times 10^{-8}$		$\times 10^{-8}$		$\times 10^{-8}$
Aluminum	+ 0.82	Glass	– 1.3	Oxygen	+ 133.6
Araldite	– 0.63	Helium	– 0.59	Perspex	– 0.5
Carbon (graphite)	– 4.4	Hydrogen	– 2.49	Polyethylene	+ 0.2
Copper	– 0.108	Lead chloride	– 0.40	P.V.C.	– 0.75
Copper sulfate	+ 7.7	Manganese chloride	+ 134	Sodium chloride	– 0.63
Ebonite	+0.75	Manganese dioxide	+ 48.3	Sulfur	– 0.62
Iron ammonium alum	+ 38.2	Manganese sulfate	+ 111	Sulfuric acid	– 0.50
Ferric hydroxide	+ 197	Mercury	– 0.21	Water	– 0.90
Ferrous sulfate	+ 52.2	Nitrogen	– 0.54		

Magnetic Properties of Some "Soft" Magnetic Materials

Alloy	Maximum Relative permeability (μ_i max)	Coercive force ($H_s/A\ m^{-2}$)	Energy loss per cycle ($E/J\ m^{-3}$)	Resistivity, ρ (ohm m)	Saturation induction (B_M/T)	Remarks
Iron, pure (total impurities < 0.005%)	200 000	4.0	30	10	2.15	Commercially impracticable
Mild steel	2 000	143	– 500	10		
Silicon iron (1.25% Si)	6 100	67.6	220			Isotropic
Silicon iron (4.25% Si)	9 000	23.9	70	60	2.0	Isotropic
Silicon iron (3% Si)	40 000	12	30	47	2.0	Anisotropic, (110).100
Silicon iron (3.8% Si)	1 400 000		< 3			Single crystal

Silicon iron (6.3% Si)	500 000		4.5				Polycrystalline, magnetically annealed: brittle
78 Permalloy (Fe21.5%Ni78.5%)	100 000	4.0				1.08	
Supermalloy (Fe16%Ni79%Mo5%)	1 000 000	0.16			60	0.79	
Ferroxcube 3 (Mn-Zn ferrite)	1 500	0.8			10^6	0.25	

Properties of Some Commerical Permanent Magnet Materials

Alloy	Composition					Remanence B_r /T	Coercivity $_BH_c$/A m^{-1}	Maximum B × H $(BH)_{max}$/ J m^{-3}	Comments
	Al	Ni	Co	Cu	Nb				
Alnico IV H	12	26	8	2		0.6	63 000	13×10^3	Isotropic
Ticonal C	8	13.5	24	3	0.6	1.26	52 000	430	Isotropic
Columax	8	13.5	24	3	0.5	1.35	64 000	64	Columnar
Pt-Co alloy			23			0.45	210 000	300	ductile
Barium Ferrite						0.2	135 000	7 550	Isotropic
(BaO · 6Fe$_3$O$_3$CO$_5$Sm)						0.85	600 000	140 000	
Elongated single domain magnet (Fe50% CO50%)						0.905	80 000	40	Mechanically weak

NOTE *The magnetic properties of materials depend critically on the manufacture and previous history of the specimen. The values in the tables above should therefore be taken as typical only.*

CHAPTER 17

Relative Humidities From Wet-and Dry-Bulb Thermometers (Exposed in Standard Screen)

The relative humidity is defined as the ratio, expressed as a percentage, of the actual vapor pressure to the saturation vapor pressure at the temperature of the dry bulb. The dry bulb thermometer is an ordinary thermometer; the "wet-bulb" thermometer is similar in design and has its bulb enclosed in a wick, the other end of which dips into water. By capillary action the thermometer bulb is wet and under the usually encountered conditions evaporation of the water lowers the temperature of the bulb. The difference in the reading of the two thermometers is the "Depression of the wet bulb." The tables below give relative humidities for various values of the dry bulb temperature and the depression. Temperatures are in degrees Celsius.

Depression of Wet Bulb/°C	Dry Bulb Temperature/°C															
	0	2	4	6	8	10	12	14	16	18	20	22	24	26	28	30
0.5	91	92	93	93	94	94	95	95	95	95	96	96	96	96	96	96
1.0	81	84	85	86	87	88	89	90	90	91	91	92	92	92	93	93
1.5	73	76	78	80	81	82	83	85	85	86	87	87	88	88	89	89
2.0	64	68	71	73	75	77	78	79	81	82	83	83	84	85	85	86
2.5	55	61	64	66	69	71	73	75	76	77	78	80	80	81	82	83
3.0	46	52	57	60	63	66	68	70	71	73	74	76	77	78	78	79
3.5	38	45	49	54	57	60	63	65	67	69	70	72	73	74	75	76
4.0	29	37	43	48	51	55	58	60	63	65	66	68	69	71	72	73
4.5	21	29	36	41	46	50	53	56	58	61	63	64	66	67	69	70
5.0	13	22	29	35	40	44	48	51	54	57	59	61	62	64	65	67
5.5	5	14	22	29	35	39	43	47	50	53	55	57	59	61	62	64
6.0		7	16	24	29	34	39	42	46	49	51	54	56	58	59	61

Depression of Wet Bulb/°C	Dry Bulb Temperature/°C															
	0	2	4	6	8	10	12	14	16	18	20	22	24	26	28	30
6.5			9	17	24	29	34	38	42	45	48	50	53	54	56	58
7.0				11	19	24	29	34	38	41	44	47	49	51	53	55
7.5				5	14	20	25	30	34	38	41	44	46	49	51	52
8.0					8	15	21	26	30	34	37	40	43	46	48	50
8.5						10	16	22	26	30	34	37	40	43	45	47
9.0						6	12	18	23	27	31	34	37	40	42	44
9.5							8	14	19	23	28	31	34	37	40	42
10.0								10	15	20	24	28	31	34	37	39

The Electromagnetic Spectrum

Type of radiation	Frequency, v Hz	Wavelength, λ m	Wave No., σ m^{-1}	Quantum Energy
	10^{24}	10^{-16}	10^{16}	12400 MeV
	10^{23}	10^{-15}	10^{15}	1240 MeV
	10^{22}	10^{-14}	10^{14}	124 MeV
gamma rays	10^{21}	10^{-13}	10^{13}	12.4 MeV
	10^{20}	10^{-12}	10^{12}	1.24 MeV
	10^{19}	10^{-11}	10^{11}	124 keV
X-rays	10^{18}	10^{-10}	10^{10}	12.4 keV
	10^{17}	10^{-9}	10^{9}	1.24 keV
	10^{16}	10^{-8}	10^{8}	124 eV
Violet	10^{15}	10^{-7}	10^{7}	12.4 eV

$\lambda \sim 4 \times 10^{-7}$m Ultra violet

Visible Spectrum

Red Infra-red
$\lambda \sim 7 \times 10^{-7}$m

Type of radiation	Frequency, v Hz	Wavelength, λ m	Wave No., σ m^{-1}
		10^{-6}	10^{6}
	10^{14}	10^{-5}	10^{5}
	10^{13}	10^{-4}	10^{4}
	10^{12}	10^{-3}	10^{3}
	10^{11}	10^{-2}	10^{2}
Microwaves, radar	10^{10}	10^{-1}	10
	10^{9}	1	1
	10^{8}	10	10^{-1}
	10^{7}	10^{2}	10^{-2}
Short waves	10^{6}	10^{3}	10^{-3}
Long waves	10^{5}	10^{4}	10^{-4}

ACOUSTIC DATA

A Speed of Sound at Room Temperature

Substance	Temp., t °C	Speed, v ms^{-1}	Substance	Speed, v ms^{-1}
Air	0	331.3	Aluminum	5100
Hydrogen	0	1284	Brass	3500
Oxygen	0	316	Copper	3800
Water	25	1498	Iron	5000
Oak (along fiber)	15	3850	Lead	1200
Glass	20	5000	Mercury	1452

N.B.—The velocity of sound can vary according to the crystalline state and previous history of the specimen. The values quoted for solids are for longitudinal waves in thin specimens.

Loudness of Sounds

Intensity in terms of threshold – intensity, I/1 min	Intensity, I/dB	Loudness, L/phon
1	0	Threshold of hearing
10	10 (1 bel)	Virtual silence
10^2	20	Quiet room
10^3	30	Watch ticking at 1 m
10^4	40	Quiet street
10^5	50	Quiet conversation
10^6	60	Quiet motor at 1 m
10^7	70	Loud conversation

10^8	80	Door slamming
10^9	90	Busy typing room
10^{10}	100	Near loud motor horn
10^{11}	110	Pneumatic drill
10^{12}	120	Near airplane engine
10^{13}	130	Threshold of pain

Limits of Audibility—Between 30 and 30 000 Hz (approximately).

Music

The consonant frequency intervals.

Name	Octave	Fifth	Fourth	Major Third	Major Sixth	Minor Third	Minor Sixth
Frequency ratio.	1 : 2	2 : 3	3 : 4	4 : 5	3 : 5	5 : 6	5 : 8

Musical scales—vibration ratios

	C	D	E	F	G	A	B	C
Basic}	24	27	30	32	36	40	45	48
Scale	1.000	1.125	1.250	1.333	1.500	1.667	1.875	2.000
Intervals		$\frac{9}{8}$	$\frac{10}{9}$	$\frac{16}{15}$	$\frac{9}{8}$	$\frac{10}{9}$	$\frac{9}{8}$	$\frac{16}{15}$

· The Basic Scale is frequently referred to as the Natural or Diatomic Scale.

The vibration numbers in the Basic Scale must bear the given ratios to each other, but their absolute values are matter of convention.

The London International Conference of May 1939 agreed that the international standard of concert pitch should be based on 440Hz for the treble A, that is, 264 for the "Middle C."

In the EQUALLY TEMPERED SCALE, the octaves remain as before, but 11 notes are introduced between them, the intervals being made equal and each $\sqrt[12]{2}$, *i.e.* 1.0595, say 1.06 (approx).

The following is such an equally tempered chromatic scale based on 440Hz as the treble A:

	Frequency, v Hz		Frequency, v Hz		Frequency, v Hz
C′	261.6	F	349.2	A	440.0
C#	277.2	F#	370.0	A#	466.2
D	293.7	G	392.0	B	493.9
D#	311.1	G#	415.3	C″	523.2
E	329.6				

Absorption Coefficients of Building Materials; Unit, Sabine

	Frequency					
	125 Hz	250 Hz	500 Hz	1000 Hz	2000 Hz	4000 Hz
Acoustic plaster, 13 mm	0.15	0.20	0.35	0.60	0.60	0.50
Acoustic tiles, 20 mm	0.10	0.35	0.70	0.75	0.65	0.50
Brick, unglazed	0.03	0.03	0.03	0.04	0.05	0.07
Carpet, on concrete	0.02	0.06	0.14	0.37	0.60	0.65
Carpet with foam underlay	0.08	0.24	0.57	0.69	0.71	0.73
Curtain, heavy velour	0.14	0.35	0.55	0.72	0.70	0.65
Linoleum, on concrete	0.02	0.03	0.03	0.03	0.03	0.02
Glass, heavy plate	0.18	0.06	0.04	0.03	0.02	0.02
Glass, window	0.35	0.25	0.18	0.12	0.07	0.04
Plaster	0.013	0.015	0.02	0 03	0.04	0.05
Plywood paneling, 10 mm	0.28	0.22	0.17	0.09	0.10	0.11
Polystyrene, expanded, 13 mm	0.05	0.15	0.40	0.35	0.20	0.20
Polyurethane foam, 50 mm	0.25	0.50	0.85	0.95	0.90	0.90
Tiles, glazed	0.01	0.01	0.01	0.01	0.02	0.02
Wood parquet	0.04	0.04	0.07	0.06	0.06	0.07

19

ASTRONOMICAL DATA

TIME

1 mean solar second = $\dfrac{1}{86400}$ of a mean solar day.

1 sidereal day = 86 l64.090 6 mean solar seconds.

1 tropical (civil) year = 365.242 mean solar days = 3.155 692 597 47 $\times 10^7$s

1 sidereal year = 365.256 mean solar days.

1 mean synodical or lunar month = 29.531 mean solar days.

N.B.—Centuries are not leap years unless divisible by 400.

DISTANCE

1 Astronomical Unit (AU) = mean sun earth distance = 1.495 985 (5) $\times 10^{11}$m

1 Parsec (pc) = 3.085 6(1) $\times 10^{16}$m = 2.062 648 $\times 10^5$ AU = 3.26l 5 ly

1 Ligth year (1y) = 9.460 5 $\times 10^{15}$m = 6.324 $\times 10^4$ AU = 0.3066 pc

THE SUN

Radius = 6.960 $\times 10^8$m = 4.326 $\times 10^5$ miles

Surface area = 6.087 $\times 10^{18}$ m^2

Volume $= 1.412 \times 10^{27}$ m^3

Mass $= 1.99 \times 10^{30}$ kg

Mean density $= 1409$ kg m^{-3}

Rate of energy production $= 3.90 \times 10^{26}$W

Gravity at surface $= 274$ ms^{-2}

Moment of inertia $= 6.0 \times 10^{46}$ kg m^2

Escape velocity at surface $= 618$ km s^{-1}

Sidereal period of rotation $= 25.38$ days

Period of rotation with respect to the earth $= 27.28$ days $\Big\}$ latitude 16°

THE MOON

Radius $= 1738$ km $= 1080$ miles

Surface area $= 3.796 \times 10^{13}$ m^2

Volume $= 2.199 \times 10^{19}$ m^3

Mass $= 7.349 \times 10^{22}$ kg $= 1/81.4 \times$ mass of earth

Mean density $= 3340$ kg m^{-3}

Sidereal period of moon about earth $= 27.32$ mean solar days

Mean synodical or lunar month $= 29.531$ mean solar days

Mean distance from the earth $= 3.844 \times 10^8$ m $= 2.39 \times 10^5$ miles

Surface area of the moon at some time visible from the earth $= 59\%$

Gravity at surface $= 1.62$ m s^{-3}

Moment of inertia $= 8.8 \times 10^{28}$ kg m^3

Escape velocity at surface $= 2.38$ km s^{-1}

The Solar System

Body	Equatorial radius, R m	Mass, M kg	Density, kg m^{-3}	Distance, d from Sun m	Surface gravity, g m s^{-2}	Ellipticity e	Eccentricity of orbit ε	Inclination to ecliptic, i degree	No. of satellites N	Sidereal period T_s	Rotational period T_r
Sun	6.980×10^8	1.989×10^{30}	1409	—	274	0	—	—	—	—	25.38d
Moon	1.738×10^6	7.353×10^{22}	3340	1.496×10^{11}	1.62	—	0.055	5.144	—	27.32d	27.32d
Mercury	2.42×10^6	3.301×10^{23}	5420	5.791×10^{10}	3.76	0	0.2056	7.004	0	87.97d	58.7d
Venus	6.085×10^6	4.869×10^{24}	5250	1.082×10^{11}	8.77	0	0.0068	3.394	0	224.7d	243d
Earth	6.378×10^6	5.978×10^{24}	5510	1.496×10^{11}	9.81	0.0034	0.0167	0	1	365.3d	23.93h
Mars	3.375×10^6	6.420×10^{23}	3960	2.279×10^{11}	3.80	0.007	0.0934	1.850	2	687d	24.6h
Jupiter	7.14×10^7	1.899×10^{27}	1330	7.783×10^{11}	24.9	0.062	0.0481	1.306	13	11.86a	9.9h
Saturn	6.04×10^7	5.685×10^{26}	680	1.427×10^{12}	10.4	0.096	0.0533	2.489	10	29.46a	10.2h
Uranus	2.36×10^7	8.686×10^{25}	1600	2.869×10^{12}	10.4	0.06	0.0507	0.773	5	84.02a	10.7h
Neptune	2.23×10^7	1.025×10^{26}	1650	4.498×10^{12}	13.8	0.02	0.0040	1.773	2	164.8a	15.8h

Notes: Ellipticity of a planet is defined by $(R_e - R_p) / R_e$, where R_e is the equatorial radius and R_p is the polar radius. The sidereal period of a planet is the time to move once round its orbit. Periods are measured in hours (h), days (d) or years (a).

CHAPTER 20

TERRESTRIAL AND GEODETIC DATA

THE EARTH

Polar radius = 6356.8 km

Equatorial radius = 6378.2 km

Mean radius = 6371 km = 3960 miles

Surface area = $5.101 \times 10^{14} \text{m}^2$

Volume = $1.083 \times 10^{21} \text{m}^3$

Mass = 5.977×10^{24} kg

Mean density = 5517 kg m^{-3}

Mean distance to the sun (AU) = 1.496×10^{11} m = 9.2868×10^7 miles

Distance to sun at perihelion = 1.471×10^{11} m = 9.136×10^7 miles

Distance to sun at aphelion = 1.521×10^{11} m = 9.447×10^7 miles

Gravity at surface = 9.80665 m s^{-2} (standard)

Moment of inertia about axis of rotation = 8.04×10^{37} kg m^2

Escape velocity at surface = 11.2 km s^{-1}

Rotational velocity at equator = 465 m s^{-1}

Mean Velocity in its orbit about the sun = 29.78 km s^{-1}

Solar constant = solar energy incident on unit area normal to the sun's rays at the earth's mean distance, per unit time = $1400 \, \text{J m}^{-2} \, \text{s}^{-1}$

$1°$ of latitude at equator = 110.5 km = 68.70 miles.

$1°$ of latitude at poles = 111.7 m = 69.41 miles.

$1°$ of longitude at equator = 111.3 km = 69.17 miles.

Inclination of equator to ecliptic = $23°27'$.

Greatest height (Mt. Everest) =8847.7 m = 29 028 ft (1954 Indian Survey).

Greatest depth (Marianas Trench) = 11 033 m = 35 960 ft.

Land area = $148.8 \times 10^6 \, \text{km}^2$ = $5.747 \times 10^7 \, \text{miles}^2$.

Ocean area = $361.3 \times 10^6 \, \text{km}^2$ = $13.95 \times 10^7 \, \text{miles}^2$.

Composition of the Atmosphere

The composition of dry air is remarkably constant all over the globe and throughout the entire troposphere. The proportions by *volume* of the various components are given below (after A.F. Paneth, 1939, 1952).

Substance	% by volume	Substance	% by volume
N_2	78.09	CH_4	2.0×10^{-4}
O_2	20.95	Kr	1×10^{-4}
Ar	0.93	H_2	5×10^{-5}
*CO_2	0.03	N_2O	5×10^{-5}
Ne	1.8×10^{-3}	Xe	9×10^{-6}
He	5.2×10^{-4}	Rn	6×10^{-18}

* *This varies somewhat near towns and industrial areas.*

The ICAO Standard Atmosphere

The International Civil Aviation Organization have defined a standard atmosphere which is an attempt to represent atmospheric conditions in temperate latitudes. At sea level, standard pressure and acceleration of gravity are assumed for a temperature of 288 K (15°C). The air is assumed to be a perfect gas of fixed composition.

Sea level properties of the ICAO atmosphere

Collision frequency	$6.9204 \times 10^9 \text{s}^{-1}$	Pressure	1.01325×10^3 Pa
Density	1.225 kg m^{-3}	Scale height	8.4344×10^3 m
Gravitational acceleration	9.80665 m s^{-2}	Speed of sound	340.29 m s^{-1}
Kinematic viscosity	$1.4607 \times 10^{-5} \text{ m}^2 \text{ s}^{-1}$	Temperature	288.15 K
Mean free path	6.6317×10^{-8} m	Thermal conductivity	2.5339×10^{-3} W m K
Molar volume	$2.3645 \times 10^{-2} \text{ m}^3/\text{mol}$		
Molecular weight	28.966	Viscosity	1.7894×10^{-5} kg/ms
Number density	$2.5475 \times 10^{25}/\text{m}^3$		
Particle speed	4.5894×10^3 m/s		

Variation of Pressure, Density, and Temperature with Height

Geometric Height, $h\,m$	Pressure, p P ascf	Density, ρ kg/m³	Temp., $T\,K$	Geometric Height, $h\,m$	Pressure, p Pa	Density, ρ kg/m³	Temp., $T\,K$
– 250	104365	1.2547	289.775	6000	47217.6	0.66011	249.187
0	101325	1.2250	288.150	7000	41105.2	0.59002	242.700
+ 250	98357.6	1.1959	286.525	8000	35651.6	0.52579	236.215
500	95461.2	1.1673	284.900	9000	30800.7	0.46706	229.733
750	92634.6	1.1392	283.276	10000	26499.9	0.41351	223.252
1000	89876.2	1.1117	281.651	15000	12111.8	0.19475	216.650
1500	84559.6	1.0581	278.402	20000	5529.3	0.08891	216.650
2000	79501.4	1.0066	275.154	25000	2594.2	0.04008	221.552
2500	74691.7	0.95695	271.906	30000	1197.0	0.01841	226.509
3000	70121.1	0.90925	268.659	32000	889.1	0.01355	228.490
3500	65780.3	0.86340	265.413	50000	80.96	1.041×10^{-3}	271
4000	61660.4	0.81935	262.166	100 000	3.095×10^{-2}	5.062×10^{-7}	213
5000	54048.2	0.73643	255.676	200 000	8.806×10^{-5}	2.56×10^{-10}	1198

NOTE *The above table is reproduced by permission of the International Civil Aviation Organization, Montreal. The last three sets of values in this table are taken from the COSPAR International Reference Atmosphere, 1965 (CIRA 1965) by permission of the publishers, North Holland Publishing Co., Amsterdam.*

Principal Elements in Earth's Crust (% by mass)

Oxygen 49.13%, Silicon 26.0%, Aluminum 7.45%, Iron 4.2%, Calcium 3.25%, Sodium 2.4%, Potassium 2.35%, Magnesium 2.35%, Hydrogen 1%, All others 1.87%.

Principal Elements in the Hydrosphere (% by mass)

Oxygen 85.89%, Hydrogen 10.82%, Chlorine 1.90%, Sodium 1.06%, All other 0.33%.

Acceleration of Gravity, (g)

At a latitude λ, and height, h (measured in meters), above sea level, the acceleration of gravity is given by the expression:

$$g/m \ s^{-2} = 9.80616 - 0.025928 \cos 2\lambda + 0.000069 \cos^2 2\lambda - 0.000003 \, h$$

Geophysical data for various places of importance.

In the following tables, values of the acceleration of gravity and the length of the seconds pendulum are calculated using the formula above. In addition, magnetic data, calculated for the year 1970 are included. These have been obtained from the International Reference Geomagnetic Field and excluding local variations should not be in error by more than 1%. Declination is positive Eastward and Angle of Dip positive downwards. Magnetic Induction for geophysical fields is often measured in *gammas.* where 1 gamma = 10^{-9} Tesla or 1 gamma = 1 nT.

Location	Position	Acceleration of Gravity, g m/s²	Length of Seconds Pendulum, l m	Declination, D Degree	Horizontal Component of Earth's Magnetic Field, H A/m¹	Horizontal Component of Earth's Magnetic Induction B_H/nT	Angle of Dip, I Degree
Equator	0°0'	9.78030	0.99094				
Madras	13°5'N 80°18'E	9.78281	0.99120	-2.23	32.4	40660	9.1
Calcutta	33°35'S 88°21'E	9.7882	0.99175	-0.82	31.2	39200	29.9
Sydney	33°55'S 151°10'E	9.7968	0.99262	11.9	20.2	25380	-64.1
Capetown	33°56'S 18°28'E	9.7966	0.99260	-24.6	9.86	12390	-65.1
Tokyo	35°40'N 139°45'E	9.79801	0.99275	-6.4	24.3	30530	48.3
New York	40°40'N 73°50'W	9.80267	0.99322	-11.6	14.6	18420	71.0
Paris	48°52'N 2°20'E	9.80943	0.99390	-5.5	16.0	20140	64.7
London	51°25'N 0°20'W	9.81183	0.99415	-7.0	15.0	18820	66.8
Edinburgh	55°57'N 3°13'W	9.8158	0.99455	-9.4	13.2	16620	70.1
Leningrad	59°55'N 30°25'E	9.81929	0.99490	7.0	12.1	15180	72.8
N'th Pole	90°0'N	9.8322	0.99621				

21

RADIOACTIVITY

The most common unit of radioactivity is the Curie (Ci). Originally defined as the volume of radon gas in equilibrium with 1 g radium, it has since become associated with the number of disintegrations occurring per second in 1 g of radium free from its daughter products viz. 3.7×10^{10} disintegrations per second. In modern usage, the curie has been redefined to agree with this result, and other units have been introduced as given below.

One *curie* (Ci) of any radioactive substance is that quantity in which 3.7×10^{10} atoms disintegrate per second. The millicurie (mCi) and microcurie (μCi) are in common usage.

The *rutherford* is the unit of activity corresponding to 10^6 disintegrations per second. Thus 37 rutherford = 1 mCi.

The *roentgen* (r) was originally suggested as a unit of radiation and has become of universal use in defining the quantities of X-rays or γ-rays present. In 1937, the Fifth International Congress of Radiobiology recommended the following definition:

The *roentgen* is the quantity of X = or γ radiation such that the associated corpuscular emission per 0.001293 g of dry air produces, in air, ions carrying 1 esu of quantity of electricity of either sign. (*N.B.* this mass of air occupies 1 cm³ at STP).

Dose rates are often measured in units of roentgen hour⁻¹ or milliroentgen hour⁻¹ (mr h⁻¹)

The *rad* is defined as the absorbed dose of radiation when 1 g of material absorbs 100 ergs of energy. 1 rad = 10^{-2} J kg⁻¹.

The *roentgen equivalent man* (rem) is the unit Dose Equivalent used in Radiation Protection. The Dose Equivalent is the product of the Absorbed Dose (measured in rad) and the quality factor Q, of the radiation. The value of Q^4 indicates how damaging the particular radiation is, compared with 200 keV X-rays. Thus, low energy β-rays have $Q = 1.7$, while neutrons impinging on the eye have $Q = 30$.

A useful, but approximate formula for calculation of dose rates from γ-ray point sources is

$$\text{Dose rate (r hr}^{-1}) \simeq (5000\ C\ E)/d^2$$

where C is the activity of the source in curies, E *is* the energy of the γ-ray emitted in MeV, and d is the distance from the source in cm. If more than one γ-ray is emitted, the total dose rate is the sum of the individual dose rates.

The naturally radioactive materials with the exception of a few isotopes, for example, ^{40}K are the heavy elements of atomic number $Z > 80$. Three "families" are known in which one substance decays to another which in turn continues the process until a stable material (lead) is attained. The decay process involves the emission of an electron (β-particle) or an α-particle from the nucleus. In the former case, the mass number, A, remains unchanged while Z increases by unity, while the latter emission involves a decrease in A of four and a decrease in Z of two as the α-particle is the helium nucleus. In any one "family" the mass numbers alter in steps of four only. In the Thorium family, each value of A can be described by the number $(4n)$, the Uranium family by $(4n + 2)$, and the Actinium family by $(4n + 3)$. The apparently missing family $(4n + 1)$ has been found as a result of the artificial production of heavy isotopes. It does not appear naturally because the longest half-life is short compared with the age of the Earth.

The Law of Radioactive Decay

All radioactive substances transform at a rate that is proportional to the number of atoms present. If there are N_0 atoms present at the zero of time, then at time, t, there are N_t, where

$$N_t = N_0 \exp{-(\lambda t)}$$

Here, λ, is a constant for the particular type of atom considered and is known as the transformation constant. The rate at which an atom decay is often measured in terms of the mean lifetime of the atom, τ, or the half-value period, T_t, which is often abbreviated to the half-life. The relation between these constants is:

$$\tau = 1/\lambda = T_t / \log_e 2$$

CHAPTER 22

PROPERTIES OF INORGANIC COMPOUNDS

In the following table, properties refer to room temperature, 293 K. Enthalpies of Formation refer to the substance in the crystalline (c), liquid (lq), or gaseous (g) states at 293 K. A negative value indicates that heat is evolved in the formation of the compound, while a positive value indicates absorption of heat. The following abbreviations are used:

bl.	black	effl.	efflorescent	s.	sublimes
col.	colorless	ex.	explodes	tetr.	tetragonal
crys.	crystals	gn.	green	trig.	trigonal
cub.	cubic	hex.	hexagonal	visc.	viscous
d.	dissociates	mono.	monoclinic	w.	white
delq.	deliquescent	rh.	rhombic	yel.	yellow

	Formula	Molecular weight, M g/mol^{-1}	Melting point, T_M K	Boiling point, T_a K	Density ρ, kg/m^3	Refractive index n	Enthalpy of formation, $\Delta H_f \theta$ kJ/mol^1	Description
Al	Al_2O_3	101.96	2290	3250	3965	1.768	– 1670 c	Corundum. w. trig.
Ag	AgBr	187.78	705	1600 (d)	6473	2.252	– 99.5 c	pale yel. cub
	AgCl	143.32	728	1820	5560	2.071	– 127 c	w. cub.
	$AgNO_3$	169.87	485	717 (d)	4352	1.744	– 123 c	col. rh.
As	$AsBr_3$	314.65	306	494	3540		– 195.0 c	col. prisms
	$AsCl_3$	181.28	265	403	2163		– 335 lq	Oily liquid
	As_2O_3	197.84	588		3738	1.755	– 1310 c	col. cub. (As_4O_6)
Au	$AuCl_3$	303.33	527 (d)		3900		– 118 c	red delq.
Ba	$BaCl_2$	208.25	1240	1820	3856	1.736	– 860.1 c	col. mono.
	BaO	153.34	2196	2300	5720	1.98	– 558.1 c	col. cub.
Be	$BeCl_2$	79.92	678	790	1899	1.719	– 511.7 c	w. delq. needles
	BeO	25.01	2800	4170	3010	1.719	– 610.9 c	w. hex.
C	CO	28.01	74	84	1.25		– 110.5 g	col. gas
	CO_2	44.01	162	195	1.98		– 393.5 g	col. gas
Ca	$CaCO_3$	100.09	1612	d	2930	1.6809	– 1206.9	Aragonite. col. rh.
	$CaCl_2$	110.99	1045	1900	2150	1.52	– 795.0 c	w. delq. cub.
	CaO	56.08	2850	3120	3300	1.837	– 635.5 c	col. cub.
Cd	$CdBr_2$	272.22	840	1136	5192		– 314.4c	w. effi. needles
	$CdCl_2$	183.32	841	1233	4047		– 389.1 c	w. cub.
	CdO	128.40	1200 (d)		8150		– 254.6 c	brown cub.
Co	$CoCl_2$	129.84	997*	1322	2940		– 326 c	blue crys. *in HCl gas
	CoO	74.93	2208		6450		– 239 c	brown cub.

	Formula	Molecular weight, M g/mol^{-1}	Melting point, T_M K	Boiling point, T_a K	Density ρ, kg/m^3	Refractive index, n	Enthalpy of formation, $\Delta H_f \theta$, kJ/mol^{-1}	Description
	Co(OH)$_2$	92.95	d		3597		– 548.9 c	rose-red rh.
Cs	CsCl	168.36	919	1560	3988	1.534	– 433.0 c	col. delq. cub.
Cu	CuO	79.54	1599		6400		– 155.2 c	bl. cub. or trig.
	CuSO$_4$	223.14			3605	1.733	– 769.9 c	gn/w. rh.

	Formula	Molecular weight, M g/mol^{-1}	Melting point, T_M K	Boiling point, T_a K	Density ρ, kg/m^3	Refractive index, n	Enthalpy of formation, $\Delta H_f \theta$, kJ/mol^{-1}	Description
	CuSO$_4$ 5H$_2$O	249.68			2284	1.537	– 2278c	blue trig.
	Cu$_2$O	143.08	1508		6000	2.705	– 166.7c	red cub.
Fe	FeS	87.91	1470	d	4740		– 95.1c	blue hex.
	Fe$_2$O$_3$	159.69	1838		5240	3.042	– 822.2c	red or bl. trig.
	Fe$_3$O$_4$	231.54	1810 (d)		5180	2.42	– 1117c	bl. cub.
H	HBr	80.92	185	206	3.5		– 36.2 g	col. gas
	HCl	36.46	158	188	1.0		– 92.3 g	col. gas
	HF	20.01	190	293	0.99		– 268.6 g	col. gas
	HI	127.91	222	238	5.66		+ 25.9 g	col. gas
	HNO$_3$	63.01	231	356	1503		– 173.2 lq	col. liquid
	H$_2$O	18.02	273	373	1000	1.333	– 285.9 lq	col. liquid
	H$_2$SO$_4$	98.08	284	610	1841		– 814.0 lq	col. visc. liquid
Hg	HgCl	236.05	670 (s)		7150	1.973	– 265* c	w. tetr. (*Hg$_2$Cl$_2$)
	HgCl$_2$	271.50	549	575	5440	1.859	– 230 c	w. rh.
	HgO	216.59	800 (d)		11100	2.5	– 90.4 c	yel. or red rh.
K	KCl	74.56	1049	1770 (s)	1984	1.490	– 435.9 c	col. cub

	KHCO$_3$	100.12	400 (d)		2170	1.482	– 959.4 c	mono.
	K$_2$CO$_3$	138.21	1164	d	2428	1.531	– 1146.1 c	w. delq.
	K$_2$O	94.20	620 (d)		2320		– 361.5 c	w. cub.
Li	LiCl	42.39	887	1600	2068	1.662	– 408.8 c	w. delq. cub.
Mg	MgBr$_2$	184.13	970		3720		– 517.6 c	w. delq.
	MgCO$_3$	84.32	620	1200	2958	1.700	– 1112 c	w. trig.
	MgCl$_2$	95.22	981	1685	2320	1.675	– 641.8 c	col. hex
	MgF$_2$	62.31	1539	2512		1.378	– 1102 c	col. tetr.
	MgH$_3$	26.33	550 (d)					w. tetr.
	MgI	278.12	1000 (d)		4430		– 359 c	w. delq.
	MgO	40.31	3100	3900	3580	1.736	– 601.8 c	col. cub.
	Mg(OH)$_2$	58.33	620		2360	1.562	– 924.7 c	w. trig.
	MgSO$_4$	120.37	1397		2660	1.56	– 1278 c	col. rh.
Mn	MnO	70.94			5440	2.16	– 385 c	gray/gn cub.
	MnO$_3$	86.94	808 (d)		5026		– 520.9 c	bl. rh.
	MnO$_3$	102.94						red delq.
	Mn$_2$O$_3$	157.87	1350 (d)		4500		– 971.1 c	brown/bl. cub.
	Mn$_2$O$_7$	221.87	279	328 (d)	2396			red oil
	Mn$_3$O$_4$	228.81	1978		4856	2.46	– 1386 c	brown/bl. tetr
N	NH$_3$	17.03	195	240	0.77		– 46.2 g	col. gas
	NH$_4$Cl	53.49	613 (s)		1527	1.64	– 315.4 c	w. cub.

	Formula	Molecular weight, M g/mol^{-1}	Melting point, T_M K	Boiling point, T_a K	Density ρ, kg/m^3	Refractive index, n	Enthalpy of formation, $\Delta H_f \theta$, kJ/mol^1	Description
	NO	30.01	110	121	1.34		+ 90.4 g	col. gas
	NO$_2$	44.01	182	185	1.98		+ 33.8 g	red/brown gas (N_2O_4)
	N$_2$O$_3$	76.01	171	277 (d)	1447		+ 83.8 g	red/brown gas
Na	NaBr	102.90	1028	1660	3203	1.641	− 359.9 c	col.cub.
	NaCl	58.44	1074	1686	2165	1.544	− 411.0 c	col.cub.
	NaF	41.99	1261	1968	2558	1.326	− 569 c	col.tetr.
	NaH	24.00	1100 (d)		920	1.470	− 57.3 c	silver needles
	NaHCO$_3$	84.00	540 (d)		2159	1.500	− 947.7 c	w. mono. powder
	NaHSO$_4$	120.06	590	d	2435		− 1126 c	col. tricl.
	NaI	149.89	924	1577	3667	1.774	− 288.0 c	col.cub.
	NaOH	40.00	592	1660	2130		− 426.7 c	w. delq.
	Na$_2$CO$_3$	105.99	1124	d	2532	1.535	− 1131 c	w. powder
	Na$_2$O	61.98	1548 (s)		2270		− 416 c	w/gray delq.
	Na$_2$SO$_4$	142.04			2680	1.477	− 1384 c	mono (→hex at 510 K)
Ni	NiCl$_2$	129.62	1274		3550		− 316 c	yel. delq.
	NiO	74.71	2260		6670	2.37	− 244 c	gm/bl. cub.
P	PCl$_3$	137.33	161	349	1574	1.503	− 320 c	col. fuming liquid
	PCl$_5$	208.24		435 (s)	4.65		− 463.2g	delq. tetr.
	PH$_3$	34.00	140	185			+ 5.2 g	col. gas
	P$_2$O$_3$	109.95	297	447*	2135		− 820 lq	w. delq. mono *in N$_2$

	P_2O_4	125.95	370	450*	2540			w. delq. rh. *in vacuo
	P_2O_5	141.94	850	875	2390		– 3012* c	w. delq. amor. *P_4O_{10}
Pb	$PbCl_2$	278.10	774	1220	5850	2.217	– 359.2 c	w. rh.
	$PbCl_4$	349.00	258	378 (ex)	3180			yel. liquid
	PbO	223.19	1161		9530		– 219.2 c	red amor.
	PbO_2	239.19	560 (d)		9375	2.229	– 276.6 c	brown tetr.
	PbS	239.25	1387		7500	3.912	– 100.4 c	lead gray cub.
	Pb_3O_4	685.57	770 (d)		9100		– 718.4 c	red amor.
Rb	RbCl	120.92	988	1660	2800	1.494	– 430.5 c	cub.
S	SO_2	64.06	200	263	2.93		– 296.9 g	col. gas
	SO_3	80.06	306	318	1927*		– 395.2 g	col. gas (*liquid)
Sb	$SbBr_3$	361.48	370	550	4148	1.74	– 260 c	col. rh.
	$SbCl_3$	228.11	347	556	3140		– 382 c	col. rh. delq.
	$SbCl_3$	299.02	276	352	2336		– 438 lq	pale yel. liquid
Si	SiC	40.10	3000		3217	2.654	– 111.7 c	blue/bl. trig.
	$SiCl_4$	169.90	203	331	1483	1.412	– 640.2 lq	col. fuming liquid
	SiH_4	32.12	88	161	1.44		+ 34 g	col. gas
	SiO	44.09	1975	2150	2130			w. cub.
	SiO_2	60.08	1880	2500	1.544		– 911 c	Quartz. hex.

Formula		Molecular weight, M g/mol^{-1}	Melting point, T_M K	Boiling point, T_a K	Density ρ, kg/m^3	Refractive index, n	Enthalpy of formation, $\Delta H_f \theta$, kJ/mol^1	Description
Sn	SnCl$_4$	260.05	240	387	2226		– 511.3 lq	cot. fuming liquid
	SnO	134.69	1350 (d)		6446		– 286 c	bl. cub.
	SnO$_2$	150.69	1400	2100 (s)	6950	1.997	– 581 c	w. tetr.
Sr	SrCl$_2$	158.53	1146	1520	3052	1.536	– 828 c	w. rh.
	SrO	103.62	2700	3300	4700	1.870	– 590 c	col. cub.
Ti	TiCl$_4$	189.71	248	409	1726		– 750 lq	col. liquid
	TiO$_2$	79.90	2098		4170	2.586	– 912 c	bl.rh.
U	UC$_2$	262.05	2650	4640	11280		– 176 c	metallic crystals
	UO$_2$	270.03	2800		10960		– 1130 c	bl. rh.
W	WC	195.86	3140	6300	15630		– 38.0 c	gray. cub. powder
	WO$_3$	231.85	1746		7160		– 840 c	yel. rh.
Zn	ZnCO$_3$	125.39	570 (d)		4398	1.818	– 813 c	w. trig.
	ZnCl$_3$	136.28	556	1005	2910	1.687	– 416 c	w. delq.
	ZnO	81.37	2100		5606	2.004	– 348 c	w. hex.

CHAPTER 23

PROPERTIES OF ORGANIC COMPOUNDS (AT 293 K)

Enthalpies of Formation refer to the substance in the crystalline (c), liquid (lq), or gaseous (g) states at 293 K. A negative value indicates the evolution of heat during the formation of the compound, while a positive value indicates absorption of heat. Enthalpy changes in combustion refer to combustion at a pressure of 1 atmosphere and temperature 293 K, the final products being liquid water, and gaseous carbon dioxide and nitrogen.

Name and formula	Molecular weight	Melting point, T_M K	Boiling point, T_B K	Density, ρ kg/m³	Refractive index, n	Enthalpy of formation, ΔH_f kJ/mol	Heat of combustion, ΔH_2 kJ/mol	Alternative name
Hydrocarbons								
Methane, CH_4	16.04	91	109			− 74.85 g	890.4 g	
Ethane, C_2H_6	30.07	90	185			− 84.7 g	1560 g	
Propane, C_3H_8	44.11	83	231			− 103.8 g	2220 g	
n-Butane, n-C_4H_{10}	58.13	135	273	579	1.3543	−146.2 lq	2877 g	
2-Methyl propane iso-C_4H_{10}	58.13	114	261	557		− 134.6 g	2869 g	Isobutane
n-Pentane, n-C_5H_{12}	72.15	143	309	626	1.3575	− 173 lq	3509 lq	
n-Hexane, n-C_6H_{14}	86.18	178	342	660	1.3751	− 198.8 lq	4163 lq	
n-Heptane, n-C_7H_{16}	100.21	183	372	638	1.3878	− 224.4 lq	4853 lq	
n-Octane, n-C_8H_{18}	114.23	216	399	702	1.3974	− 250 lq	5512 lq	

Name and formula	Molecular weight	Melting point, T_M K	Boiling point, T_B K	Density, ρ kg/m³	Refractive index, n	Enthalpy of formation, $\Delta H_f \theta$ kJ/mol	Heat of combustion, ΔH_2 kJ/mol	Alternative Name
Ethene, n-C_3H_4	28.05	104	169	1.26		+ 52.3 g	1411 g	Ethylene
Propene, C_3H_6	42.08	88	226	519	1.3567	+ 20.4 g	2059 g	Propylene
Ethyne, C_2H_2	26.04	192	189	618		+ 226.7 g	1300 g	Acetylene
Benzene, C_6H_6	78.12	279	353	879	1.5011	+ 48.7 lq	3273 lq	
Cyclohexane, C_6H_{12}	84.16	280	354	779	1.4266	− 156.2 lq	3924 lq	

Halogen derivatives of

Hydrocarbons

Monochloro-methane, CH_3Cl	50.49	175	249	916		– 81.9 g	687 g	Methyl chloride
Dichloromethane, CH_2Cl_2	84.93	178	313	1327	1.4242	– 117 lq	447 g	Methylene dichloride
Trichloromethane, $CHCl_3$	119.38	210	335	1483	1.4459	– 132 lq	373 lq	Chloroform
Tetrachloromethane, CCl_4	153.82	250	350	1594	1.4601	–139.5 lq	156 lq	Cargon tetrachloride
Bromomethane, CH_3Br	94.94	180	277	1676	1.4218	– 35.6 g	770 g	Methyl bromide
Iodomethane, CH_3I	141.94	207	316	2279	1.5380	– 8.4 lq	815 lq	Methyl iodine

Alcohols

Methanol, CH_3OH	32.04	179	338	791	1.3288	– 238.7 lq	726.5 lq	
Ethanol, C_2H_5OH	46.07	156	352	789	1.3611	– 277.7 lq	1371 lq	
n-Propanol, $n-C_3H_7OH$	60.11	147	371	803	1.3850	– 300 lq	2017 lq	
Propane-1,2,3-triol, $C_3H_8O_3$	92.11	293	d	1261	1.4746	– 103.9 lq	1661 lq	Glyerol

Acids

Ethanoic acid, CH_3COOH	60.05	290	391	1049	1.3716	– 488.3 lq	876 lq	Acetic acid
Propanoic acid, C_2H_5COOH	74.08	252	414	993	1.3869	– 509 lq	1527 lq	Propionic acid
n-Butanoic acid, $n-C_3H_7COOH$	88.12	269	437	958	1.3980	– 538.9 lq	2194 lq	
Benzoic acid, C_6H_5COOH	122.13	396	522	1266	1.504	– 390 c	3227 c	

Miscellaneous

Ethanal, CH_3CHO	44.05	152	294	783	1.3316	– 166.4 g	1192 lq	Acetaldehyde
2-Propanone, $CH_3 \bullet CO \bullet CH_3$	58.08	178	329	790	1.3588	– 216.7 g	1821 g	Acetone
Methoxymethane, $CH_3 \bullet O \bullet CH_3$	46.07	135	250			– 185 g	1454 g	Dimethylether
Ethoxyethane, $C_2H_5 \bullet O \bullet C_2H_5$	74.12	157	308	714	1.3526	– 279.6 lq	2761 lq	Diethylether
Urea, $CO(NH_2)_2$	60.06	408	d	1323	1.484	– 333.2 c	634 c	
Glycine, $NH_2 \bullet CH_2 \bullet COOH$	75.07	d		828		– 528.6 c	981 c	

CHAPTER 24

THE FUNDAMENTAL CONSTANTS

Certain physical constants have special importance on account of their universality or place in fundamental theory. These are given below, first in SI and then in cgs units.

The figure in brackets which follows the final digit, is the estimated uncertainty in the last digit.

Thus $c = 2.997\ 924\ 590(8) \times 10^8 \text{m s}^{-1}$ could be written

$c = (2.997\ 924\ 590 \pm 0.000\ 000\ 008) \times 10^8 \text{ m s}^{-1}$.

	Symbol	Quantity	Value	Multiplier and units SI
General contents	C	Speed of light in vacuo	2.997 924 590 (8)	$10^8 ms^{-1}$
	μ_0	Permeability of free space	4π	10^{-7} H m^{-1}
	ε_0	Permittivity of free space	8.854 19 (1)	10^{-12} F m^{-1}
	e	Elementary charge	1.602 192 (7)	10^{-19} C
			or 4.803 25(2)	—
	h	Planck's constant	6.626 20(5)	10^{-34} J s
	$h/2\pi$		1.054 592(8)	10^{-34} J s
	h/e	Quantum charge ratio	4.135 708(14)	10^{-15} J s C^{-1}
			or 1.379 523(5)	—
	α	Fine structure	7.297 351(11)	10^{-3}
		$\text{constant} = \dfrac{e^2}{2h\varepsilon_0 c}$		
	$1/\alpha$		1.370 360(2)	10^2
	G	Gravitational constant	6.673(3)	10^{-11} N m^2 kg^{-2}
	Z_0	Impedance of free space	3.767 304(1)	10^2 Ohm
Electron	m_e	Electron rest mass	9.109 56(5)	10^{-31} kg
	$m_e c^2$	Electron rest energy	8.187 26(6)	10^{-14} J
			or 5.110 041(16)	10^{-1} Me V
	e/m_e	Electron change mass ratio	1.758 803(5)	10^{11} C kg^{-1}
			or 5.272 759(16)	—
	λ_c	Compton wave length of electron	2.426 310(7)	10^{-12} m
	r_c	Classical radius of electron	2.817 939(13)	10^{-15} m
Proton	m_p	Proton rest mass	1.672 614(11)	10^{-27} kg
	$m_p C^2$	Proton rest energy	1.503 271(15)	10^{-10} J
			or 9.382 59(5)	10^2 MeV
	e/m_p	Proton charge-mass ratio	9.578 97(11)	10^7 C kg^{-1}
			or 2.871 70(3)	—
	λ_{cp}	Proton Compton wavelength	1.321 441(9)	10^{-15} m
	γ	Gyromagnetic ratio	2.675 197(8)	10^8 S^{-1} T^{-1}
	e^γ	Gyromagnetic ratio (uncorrected for diamagnetism	2.675 127(8)	10^8 S^{-1} T^{-1}
Neutron	m_n	Neutron rest mas	1.674 920(11)	10^{-27} kg
	$m_n c^2$	Neutron rest energy	1.505 353(15)	10^{-10} J
			or 9.395 53(5)	10^2 MeV

Atomic constants	R	Rydberg constant	1.097 3731(1)	10^7m^{-1}
	a_0	Bohr radius	5.291 772(8)	10^{-11} m
	μ	Bohr magneton	9.274 10(6)	10^{-24} J T^{-1}
	μ_0	Nuclear magneton	5.050 95(5)	10^{-27} J T^{-1}
	μ/hc	Zeeman splitting constant	4.668 60(7)	10^1 m^{-1} T^{-1}
Matter in bulk	N	Avogadro constant	6.022 17(4)	10^{23} mol^{-1}
			or 6.022 17(4)	10^{26} kg mol^{-1}
	F	Faraday	9.648 67(5)	10^4 C mol^{-1}
			or 2.892 599(16)	—
	V_0	Normal volume of perfect gas	2.241 36(30)	$10^{-2}\text{m}^3/\text{mol}$
	R	Gas constant	8.314 3(3)	10^0 J/K mol
	k	Boltzmann constant	1.380 62(6)	10^{-23} J/K
	σ	Stefan's constant	5.669 9(9)	10^{-8} W/m^2 K^4